Shape Memory Alloys - New Advances

Edited by
Mohammad Asaduzzaman Chowdhury
and Mohammed Muzibur Rahman

Published in London, United Kingdom

Shape Memory Alloys - New Advances
http://dx.doi.org/10.5772/intechopen.111044
Edited by Mohammad Asaduzzaman Chowdhury and Mohammed Muzibur Rahman

Contributors
Gianluca Rizzello, Hasani Chauke, Md. Hosne Mobarak, Mohammad Asaduzzaman Chowdhury, Nayem
Hossain, Pankaj Biswas, Paul Motzki, Phuti Ngoepe, Ramogohlo Diale, Susmita Datta, Takuya Taniguchi

Notice

Statements and opinions expressed in the chapters are these of the individual contributors and not
necessarily those of the editors or publisher. No responsibility is accepted for the accuracy of
information contained in the published chapters. The publisher assumes no responsibility for any
damage or injury to persons or property arising out of the use of any materials, instructions, methods
or ideas contained in the book.

First published in London, United Kingdom, 2024 by IntechOpen
IntechOpen is the global imprint of INTECHOPEN LIMITED, registered in England and Wales,
registration number: 11086078, 5 Princes Gate Court, London, SW7 2QJ, United Kingdom

British Library Cataloguing-in-Publication Data
A catalogue record for this book is available from the British Library

Additional hard and PDF copies can be obtained from orders@intechopen.com

Shape Memory Alloys - New Advances
Edited by Mohammad Asaduzzaman Chowdhury and Mohammed Muzibur Rahman
p. cm.
Print ISBN 978-1-83769-727-4
Online ISBN 978-1-83769-726-7
eBook (PDF) ISBN 978-1-83769-728-1

We are IntechOpen,
the world's leading publisher of
Open Access books
Built by scientists, for scientists

6,900+
Open access books available

185,000+
International authors and editors

200M+
Downloads

Our authors are among the

156
Countries delivered to

Top 1%
most cited scientists

12.2%
Contributors from top 500 universities

Interested in publishing with us?
Contact book.department@intechopen.com

Numbers displayed above are based on latest data collected.
For more information visit www.intechopen.com

Meet the editors

Mohammad Asaduzzaman Chowdhury is a Professor of Mechanical Engineering at Dhaka University of Engineering and Technology (DUET), Bangladesh. His research interests include advanced materials, nanotechnology, energy storage materials, semiconductors, next-generation batteries, 2D materials, engineering tribology, surface engineering, coating technology, and biomedical engineering. He has served as a keynote speaker, session chair, editorial board member, and reviewer for reputable journals and conferences. He is an author and editor of more than ten books. Additionally, he has published more than 220 research and review articles in international journals and conference proceedings. He is a consultant, advisor, and expert member of many government and autonomous organizations. He has 23 years of teaching and research experience. He is a member of the American Society of Mechanical Engineering (ASME) and the Physical Society of Bangladesh. He is also a fellow of the Institution of Engineers, Bangladesh. He has participated in various cultural and social activities. He has committed to contributing articles, stories, lyrics, and poems to various newspapers and pertinent publications. Dr. Chowdhury is listed among the top 2% of all scientists worldwide. His tribology publications have significantly contributed to the reduction of energy consumption, and his novel borophene synthesis process, which is temperature-dependent and ultrasound-assisted, has opened up a new avenue for nanoelectronics and optoelectronics research.

Mohammed Muzibur Rahman received his BSc and MSc from Shahjalal University of Science and Technology, Bangladesh, in 1999 and 2001, respectively. He received his Ph.D. from Chonbuk National University, South Korea, in 2007. After his Ph.D., Dr. Rahman completed a postdoctoral fellowship and worked as an assistant professor at pioneer research centers and universities located in South Korea, Japan, and Saudi Arabia. Presently, he is a professor at the Center of Excellence for Advanced Materials Research (CEAMR) and Chemistry Department at King Abdulaziz University, Saudi Arabia. He has published more than 550 research articles, 11 US patents, and 42 book chapters. He has edited twenty-two books and attended several international and domestic conferences. His research interests include carbon nanotubes, sensors, nanotechnology, nanocomposites, nanomaterials, nanoparticles, carbon nanofibers, photocatalysis, semiconductors, electrocatalysis, ionic liquids, surface chemistry, electrochemistry, and more.

Contents

Preface

This edited volume is a collection of reviewed and relevant research chapters on the developments in shape memory alloys. It includes scholarly contributions by various authors and is edited by a group of experts in materials science and technology. Each contribution comes as a separate chapter, complete in itself but directly related to the book's topics and objectives.

The book is organized into two sections: "Introduction to Shape Memory Alloys" and "Design and Development of Shape Memory Alloys".

Section 1 includes Chapter 1, "Introductory Chapter: Introduction to Shape Memory Alloys", and Chapter 2, "Smart Shape Memory Alloy Actuator Systems and Applications". Section 2 includes Chapter 3, "Design and Development of B2 $Ti_{50}Pd_{50-x}M_x$ (M = Os, Ru, Co) as Potential High-Temperature Shape Memory Alloys", Chapter 4, "Superelastic Behaviors of Molecular Crystals", and Chapter 5, "A Review on Present Status of Friction Stir Welding of NiTinol, a Functionally Advanced, Versatile and Widely Used Shape Memory Alloy".

We are grateful to Dhaka University of Engineering & Technology (DUET), Bangladesh and King Abdulaziz University, Saudi Arabia for facilitating our participation in this work.

We greatly benefited from the support that was provided by Publishing Process Manager Ms. Zrinka Tomicic and the publisher IntechOpen. We express our gratitude to the authors for their valuable contributions.

Mohammad Asaduzzaman Chowdhury
Dhaka University of Engineering and Technology,
Gazipur, Bangladesh

Mohammed Muzibur Rahman
King Abdulaziz University,
Jeddah, Saudi Arabia

Section 1

Introduction to Shape Memory Alloys

Chapter 1

Introductory Chapter: Introduction to Shape Memory Alloys

Mohammad Asaduzzaman Chowdhury, Nayem Hossain and Md. Hosne Mobarak

1. Introduction

There are a variety of applications for shape memory alloys (SMAs), including improvements to soft robotics and robotic actuation [1]. Additionally, as a result of their work, adaptive materials, better alloys, and miniaturization techniques for MEMS and nanoscale devices have been developed [2, 3]. The design is made better by the improved modeling. The promotion of multifunctionality is enabled by the combination of SMAs and smart materials [4]. In medicine, they are used for surgical instruments, stents, and orthodontics, and they have contributed to medical advancements such as self-expanding stents and smart materials [5]. SMAs are used in the aerospace and automotive industries to improve the performance and safety of airplanes and vehicles [6, 7]. Additionally, there is evidence that they could capture energy from fluctuations in temperature [8]. SMAs play an essential role in various fields, including healthcare, transportation, and the development of energy solutions, among others.

2. Emerging trends

SMAs are utilized in soft robotics and other robotic applications. Their ability to produce precise and repeatable motion with low noise and excellent energy economy makes them appropriate for actuation in many robot designs [9]. Additionally, advances in SMA research have led to the development of adaptive materials for use in numerous industries [10]. These materials can change their properties, such as stiffness or shape, in reaction to external stimuli, making them attractive for applications like self-healing materials and adaptive constructions [11, 12]. Furthermore, research has continued to develop SMA alloy compositions to better their qualities. This includes the creation of novel alloys with superior performance features, such as increased recoverability [13].

SMAs have been the focus of miniaturization efforts by researchers, with the end goal being their use in microelectromechanical systems (MEMS) and nanoscale devices. These extremely small SMAs have the potential to be used in a variety of applications, including medical devices, sensors, and other kinds of small-scale systems [14, 15]. In addition, developments in modeling and simulation methods have led to a greater understanding of the behavior of SMAs, which has been made possible by these techniques. Because of this, SMA-based system design and optimization have become significantly more precise [16].

IntechOpen

The integration of shape-memory alloys (SMAs) with other intelligent materials, such as piezoelectric and shape-memory polymers, has resulted in the creation of new prospects for the development of multifunctional materials and devices [17]. In addition, numerous businesses have developed goods and solutions based on SMAs, which has led to an increase in the availability of these items in the commercial market. This includes actuators, sensors, and other components based on SMA that can be utilized in a variety of applications [18].

The use of shape memory alloys, often known as SMAs, has been noted to be increasing in the medical industry, particularly in the realm of minimally invasive therapies and implanted devices. In this context, SMAs are being utilized more and more in the design of orthodontic appliances, stents, and surgical equipment [19, 20]. These materials, which are renowned for their exceptional properties, facilitate the development of cutting-edge solutions such as self-expanding stents, which can autonomously adapt to vessel constraints, and smart materials that are highly responsive to bodily cues, such as fluctuations in temperature. This heralds a new era in patient-centered, precision healthcare interventions [21, 22].

Shape memory alloys (SMAs) have emerged as crucial materials in the aerospace and automotive industries, where they play a significant role in enhancing the performance and safety of essential components. These versatile materials are utilized in variable geometry components, such as aircraft flaps, landing gear, and engine parts [23, 24]. The remarkable ability of these entities to undergo changes in shape in response to external factors, such as fluctuations in temperature or mechanical strain, has been of great significance [25]. The ability to adapt facilitates more effective and secure functioning of aircraft and vehicles, hence making significant contributions to advancements in both industries through the enhancement of aerodynamic characteristics, fuel efficiency, and the overall dependability of crucial components [26].

Researchers have been diving into the creative application of shape memory alloys (SMAs) for energy harvesting, leveraging their unique ability to turn shape changes into mechanical work [27]. This pioneering concept involves inserting SMAs into constructions exposed to regular temperature variations, such as bridges and buildings, as a means of collecting otherwise lost energy [28]. Through this innovative integration, these flexible materials may generate mechanical effort by reacting to temperature differences, effectively serving as a sustainable energy source [29]. This development opens possibilities for powering a range of devices and sensors, delivering a sustainable solution to the growing energy demands in our infrastructure.

Shape memory alloys (SMAs) have emerged as important assets in the aerospace and automotive sectors, where they play a pivotal role in improving the performance and safety requirements of mission-critical components. These adaptable materials are readily integrated into adjustable geometry parts, including but not limited to aircraft flaps, landing gear, and engine components. Their amazing capacity to morph in response to external stimuli, whether owing to temperature changes or mechanical stresses, has shown to be an invaluable trait. This malleability paves the way for more effective and safe airplane and automotive operations, delivering advancements in aerodynamic features, fuel efficiency, and the overall dependability of these vital systems, ultimately determining the future of transportation technologies [30–33].

Author details

Mohammad Asaduzzaman Chowdhury[1*], Nayem Hossain[2] and Md. Hosne Mobarak[2]

1 Department of Mechanical Engineering, Dhaka University of Engineering and Technology (DUET), Gazipur, Bangladesh

2 Department of Mechanical Engineering, IUBAT-International University of Business Agriculture and Technology, Dhaka, Bangladesh

*Address all correspondence to: asadzmn2014@yahoo.com

IntechOpen

References

[1] Ruth DJS, Sohn JW, Dhanalakshmi K, Choi SB. Control aspects of shape memory alloys in robotics applications: A review over the last decade. Sensors. 2022;**22**(13):4860

[2] Chaudhari R, Vora JJ, Parikh DM. A review on applications of nitinol shape memory alloy. Recent Advances in Mechanical Infrastructure: Proceedings of ICRAM. 2021;**2020**:123-132

[3] Orlov AP, Frolov AV, Lega PV, Kartsev A, Zybtsev SG, Pokrovskii VY, et al. Shape memory effect nanotools for nano-creation: Examples of nanowire-based devices with charge density waves. Nanotechnology. 2021;**32**(49):49LT01

[4] Gopalakrishnan T, Chandrasekaran M, Saravanan R, Murugan P. An ample review on compatibility and competence of shape memory alloys for enhancing composites. Advances in Materials Science and Engineering. 2022;**2022**:6988731

[5] Nair VS, Nachimuthu R. The role of NiTi shape memory alloys in quality of life improvement through medical advancements: A comprehensive review. Proceedings of the Institution of Mechanical Engineers, Part H: Journal of Engineering in Medicine. 2022;**236**(7):923-950

[6] Rajput GS, Vora J, Prajapati P, Chaudhari R. Areas of recent developments for shape memory alloy: A review. Materials Today: Proceedings. 2022;**62**:7194-7198

[7] Shreekrishna S, Nachimuthu R, Nair VS. A review on shape memory alloys and their prominence in automotive technology. Journal of Intelligent Material Systems and Structures. 2023;**34**(5):499-524

[8] Adeodato A, Duarte BT, Monteiro LLS, Pacheco PMC, Savi MA. Synergistic use of piezoelectric and shape memory alloy elements for vibration-based energy harvesting. International Journal of Mechanical Sciences. 2021;**194**:106206

[9] Patterson ZJ, Sabelhaus AP, Majidi C. Robust control of a multi-axis shape memory alloy-driven soft manipulator. IEEE Robotics and Automation Letters. 2022;**7**(2):2210-2217

[10] Patadiya J, Gawande A, Joshi G, Kandasubramanian B. Additive manufacturing of shape memory polymer composites for futuristic technology. Industrial & Engineering Chemistry Research. 2021;**60**(44):15885-15912

[11] Chen W, Lin B, Feng K, Cui S, Zhang D. Effect of shape memory alloy fiber content and preloading level on the self-healing properties of smart cementitious composite (SMA-ECC). Construction and Building Materials. 2022;**341**:127797

[12] Taheri-Boroujeni M, Ashrafi MJ. Self-healing performance of a microcapsule-based structure reinforced with pre-strained shape memory alloy wires: 3-D FEM/XFEM modeling. Journal of Intelligent Material Systems and Structures. 2023;**34**:2192-2206. DOI: 1045389X231170163

[13] Cai WS, Chen T, Lu HZ, Ma HW, Liu Z, Yan A, et al. Achieving high strength and large recoverable strain by designing honeycomb-structural dual-shape-memory-alloy composite. Materials Science and Engineering: A. 2023;**886**:145722

[14] Hmede R, Chapelle F, Lapusta Y. Review of neural network modeling

of shape memory alloys. Sensors. 2022;**22**(15):5610

[15] Holman H, Kavarana MN, Rajab TK. Smart materials in cardiovascular implants: Shape memory alloys and shape memory polymers. Artificial Organs. 2021;**45**(5):454-463

[16] Yedla N, Salman SA, Karthik V. Molecular dynamics simulations for nanoscale insight into the phase transformation and deformation behavior of shape-memory materials. Shape memory composites based on polymers and metals for 4D printing. Processes, Applications and Challenges. 2022:67-80

[17] Sukumaran S, Chatbouri S, Muslum G, Rouxel D, Zineb TB. Hybrid composites with shape memory alloys and piezoelectric thin layers. In: Engineered Polymer Nanocomposites for Energy Harvesting Applications. Nederlands: Elsevier; 2022. pp. 225-265

[18] Balasubramanian M, Srimath R, Vignesh L, Rajesh S. Application of shape memory alloys in engineering–A review. Journal of Physics: Conference Series. 2021;**2054**(1):012078

[19] Mohammed SH, Shahatha SH. Shape memory alloys, properties and applications: A review. In: AIP Conference Proceedings. Vol. 2593. USA: AIP Publishing; 2023

[20] Dengiz D, Goldbeck H, Curtis SM, Bumke L, Jetter J, Quandt E. Shape memory alloy thin film auxetic structures. Advanced Materials Technologies. 2023;**2201991**

[21] Todorov TS, Fursov AS, Mitrev RP, Fomichev VV, Valtchev SS, Il'in AV. Energy harvesting with thermally induced vibrations in shape memory alloys by a constant temperature heater. IEEE/ASME Transactions on Mechatronics. 2021;**27**(1):475-484

[22] Phillips JW, Prominski A, Tian B. Recent advances in materials and applications for bioelectronic and biorobotic systems. Viewpoints. 2022;**3**(3):20200157

[23] Singh S, Resnina N, Belyaev S, Jinoop AN, Shukla A, Palani IA, et al. Investigations on NiTi shape memory alloy thin wall structures through laser marking assisted wire arc based additive manufacturing. Journal of Manufacturing Processes. 2021;**66**:70-80

[24] Shah PN, Blades EL, Nucci MR, Reveles ND, Turner TL, Lockard DP. Fully Coupled Aeroelastic Stability Analysis of Adaptive Shape Memory Alloy Structural Technologies for Airframe Noise Reduction. USA: NASA; 2023

[25] Dezaki ML, Bodaghi M, Serjouei A, Afazov S, Zolfagharian A. Adaptive reversible composite-based shape memory alloy soft actuators. Sensors and Actuators A: Physical. 2022;**345**:113779

[26] Khan S, Pydi YS, Prabu SM, Palani IA, Singh P. Development and actuation analysis of shape memory alloy reinforced composite fin for aerodynamic application. Sensors and Actuators A: Physical. 2021;**331**:113012

[27] Dauksher R, Patterson Z, Majidi C. Characterization and analysis of a flexural shape memory alloy actuator. Actuators. 2021;**10**(8):202

[28] Mirzaey E, Shaikh MR, Rasheed M, Ughade A, Khan HA, Shaw SK. Shape memory alloy reinforcement for strengthening of RCC structures—A critical review. Materials Today: Proceedings. 2023;**13**:1801

[29] Ntina MI, Efthymiou E, Sophianopoulos DS. Structural applications of shape memory alloys for seismic resilience enhancement. In: Proceedings of the 10th International Conference on Behaviour of Steel Structures in Seismic Areas. Greece: University of Thessaly Institutional Repository; 2022

[30] Riccio A, Sellitto A, Ameduri S, Concilio A, Arena M. Shape memory alloys (SMA) for automotive applications and challenges. Shape Memory Alloy Engineering. 2021:785-808

[31] Ferede E, Karakalas A, Gandhi F, Lagoudas DC. Numerical investigation of autonomous camber morphing of a helicopter rotor blade using shape memory alloys. In: Proceedings of the 77th Annual Vertical Flight Society Forum and Technology Display. USA: Rensselaer University; 2021

[32] Guo Y. The Applications of Shape Memory Materials in Modern car Industry and Future Trends. Finland: LUT University; 2023

[33] Ozair HUMA, Khurram AA, Baluch AUH, Wadood ABDUL, Qazi IBRAHIM. Shape memory hybrid composites for aerospace applications. In: Materials Science Forum. Vol. 1068. Switzerland: Trans Tech Publications Ltd.; 2022. pp. 93-100

Chapter 2

Smart Shape Memory Alloy Actuator Systems and Applications

Paul Motzki and Gianluca Rizzello

Abstract

Shape memory alloys (SMAs) have been established in a wide range of applications. Lead by the medical sector, nickel-titanium-based alloys are used for the realization of stents, guide wires, and other medical or surgical equipment. Besides this field, where mainly the superelastic material characteristics are used, first products based on actuation *via* the shape memory effect have been introduced to the market. These shape memory actuators or actuator systems either use temperature change directly for the realization of work output, for example, in thermostat valves, or are thermally activated by an applied electric current. This chapter gives an overview of recent SMA-based actuator systems and applications in a variety of fields from industry over bionics to automotive and aerospace.

Keywords: shape memory alloys, SMA, smart materials, artificial muscles, actuators

1. Introduction

Thermal shape memory alloys are materials that can undergo a reversible phase transformation between a high-temperature phase austenite and a low-temperature phase martensite. Geometrical changes occurring during this phase transformation are associated with high forces, so that these materials are capable of performing mechanical work. This chapter is oriented on the chapter *"Thermische Formgedächtnislegierungen"* in the German book publication *"Smart Materials"* [1].

The first documented observations of the shape memory effect date back to 1932, when the Swedish physicist Arne Ölander discovered a "rubbery" behavior of an Au-47.5Cd alloy [2]. The term "shape memory effect" (SME) was first introduced by Chang and Read in 1951 [3, 4]. In 1963, the U.S. Naval Ordinance Laboratory published the first studies of the SME in binary nickel-titanium alloys [5]. Nickel-titanium alloys, also known as "nitinol," NiTi or Ni-Ti, represent the most popular shape memory alloy (SMA) nowadays due to their attractive mechanical properties and corrosion resistance [6, 7]. In the medical sector, the so-called "superelastic" SMAs are widely used, but also in the field of small actuator systems and microactuators, SMA-based applications see a rapidly growing interest [8, 9].

2. Shape memory effect

The thermal SME describes the ability of a material to "remember" its original shape and to return to this initial shape after severe deformation by a reversible phase transformation [10–12]. This effect is mainly observed in metallic alloys, but also occasionally in ceramics (e.g., ZrO2) and polymers (e.g., PTFE) [13].

The two prominent phases martensite and austenite occurring in thermal SMAs only denote a microstructure type and do not allow any conclusion on the chemical composition of the material. In this context, martensite refers to a crystalline structure formed only by shear distortion of the lattice and not by diffusion. This shear distortion is called martensitic transformation (MT). Since the atomic bonds are not broken during MT, the local chemical composition remains unchanged [14]. To initiate MT, the austenite phase must be destabilized. This can be done by increasing the stress or reducing the temperature. The resulting thermodynamic imbalance forces a shear distortion, while the volume of the cell remains constant [15]. In the case of temperature-induced martensite formation in a stress-free state, the change in shape is negligible. Consequently, the local distortions are opposite, so that the macroscopic effect cancels out. The martensitic crystals of different orientation are called variants. If two variants face each other, they are called twins. Under the influence of mechanical stress, the boundaries of the twins shift, with some variants aligning along the mechanical stress field. This unbalances the random variant distribution and preferential variants are formed. This process is called de-twinning and results in a macroscopic shape change. As a result, a shape memory alloy can have three different microstructural states: Austenite (Aus), twinned martensite, and de-twinned martensite.

The popular nickel-titanium alloy has a body-centered cubic crystal lattice in its austenitic phase. In contrast, the lattice in martensite is monoclinic. The two possible orientations of this monoclinic structure in twinned martensite are also referred to as martensite plus (M^+) and martensite minus (M^-).

2.1 One-way effect

A body made of a shape memory alloy has the identical shape in the stress-free state after cooling as in austenite. However, the martensitic microstructure shows a comparatively large yielding due to the movable twin boundaries (**Figure 1**). When the stress reaches σ_s (1), the martensite variants begin to align in the direction of the force. When the stress reaches σ_f, all variants are aligned in the direction of force and the martensite is in a completely de-twinned form (2). Further loading of the material leads to further elastic deformation until the yield point is reached. When the material is unloaded, it remains in the de-twinned state and retains a macroscopic change in shape (3). Since this change in shape is due to the movement of twin boundaries and not to lattice defects, this behavior is referred to as pseudoplasticity. The temperature-induced martensitic transformation starts when the austenite start temperature (A_{s0}) is reached (4). Since the martensitic transformation is fully reversible, the distortion state of the de-twinned martensite is equalized by the resulting austenitic microstructure. The original orientation of the martensite variants is irrelevant. As a result, the shape change applied during de-twinning is completely reset (one-way effect). The material returns to its original shape, it has a shape memory. This state is reached when the temperature rises to the value of the austenite finish temperature (A_{f0}). In the interval between $A_{s0} < T < A_{f0}$, part of the thermal energy supplied is converted

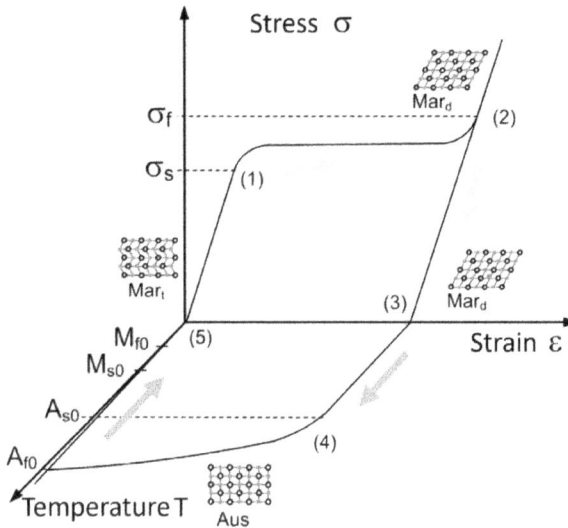

Figure 1.
Schematic of the one-way effect in a σ(ε, T) diagram according to Ref. [15].

into mechanical work. This part can be used in actuators. However, to enable cyclic activation, the material must undergo a renewed martensitic transformation. This transformation requires a driving temperature difference and starts at M_{s0}. The value of this temperature difference depends on the alloy and is between 20 and 40 K for common SMAs. The cause of the hysteresis is the friction between the twin boundaries.

2.2 Two-way effect

Both the manufacturing process (cold drawing) and repeated thermo-mechanical cycling can specifically generate internal stresses in the material. The cause of these internal stresses are dislocations and precipitations resulting from the thermo-mechanical stress. As a result, the transformation of austenite into martensite (cooling) leads to the formation of martensite variants that are energetically favorably oriented to the internal stresses. This means that no twin structure is formed in the microstructure during the transformation, but the material transforms directly into de-twinned martensite [16]. Macroscopically, this means that a change in shape also occurs during cooling. The material changes between a low-temperature form and a high-temperature form. The so-called intrinsic two-way effect is thus triggered by exclusively thermally induced phase transformations. The schematic representation of the two-way effect can be found in **Figure 2**.

The microstructural defects already occur at thermo-mechanical loads during normal actuator operation of the SMA [17]. This means that in a shape memory actuator, the so-called training effect sets in during the first cycles. As a result, there is a change in the stress-strain behavior as well as the formation of a two-way effect. Therefore, it is necessary to cycle the material sufficiently before characterization or before using it as actuator material [18]. After a sufficient number of load cycles, a stable state develops.

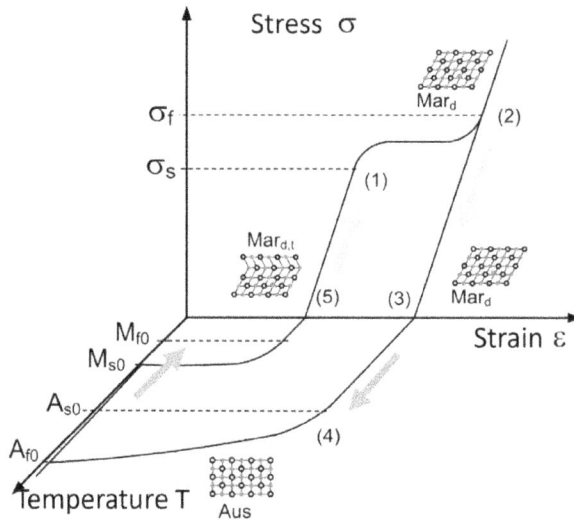

Figure 2.
Schematic of the intrinsic two-way effect in a $\sigma(\varepsilon, T)$ diagram according to Ref. [15].

The extrinsic two-way effect can be described as a continuous repetition of the one-way effect. Here, an external load, such as a mass or a spring, ensures that the SMA material is always deformed back to its martensitic phase. Thus, heating and cooling generate a transformation between two defined states in the respective phase (**Figure 3**).

At point (1), the SMA is in twinned martensite. An external load leads first to an elastic deformation (2) and then, due to de-twinning in the martensite, to a "quasi- or pseudoplastic" deformation in point (3). If the material is now heated, the transformation to austenite begins at point (4) and is completed at initial strain at point (5). On subsequent cooling, the material returns directly to the de-twinned state at point (3) due to the external load, so that repeated setting of states (3) and (5) is possible by thermal activation and cooling. This effect is primarily used in actuator technology.

3. Shape memory alloy actuators

In order to use the shape memory effect in actuator systems, the material must undergo the temperature induces phase transformation to austenite from a deformed martensitic state. Here, the temperature of the material is the external control variable. While the mechanical stress results according to the mechanical boundary condition, the displacement depends on how much the material was de-twinned before activation. To use the maximum strain, the preload of the actuator must be greater than the stress required for complete de-twinning. If this condition cannot be met by the load to be moved, a restoring element is necessary. Here, the mechanical spring in particular has become established in practical applications. This offers the advantage of a simple design, a system integration that is usually easy to represent, and automatic resetting as soon as the temperature drops below the martensite start temperature. However, if a position is to be maintained with such an actuator system, the SMA material must permanently have a temperature above martensite start

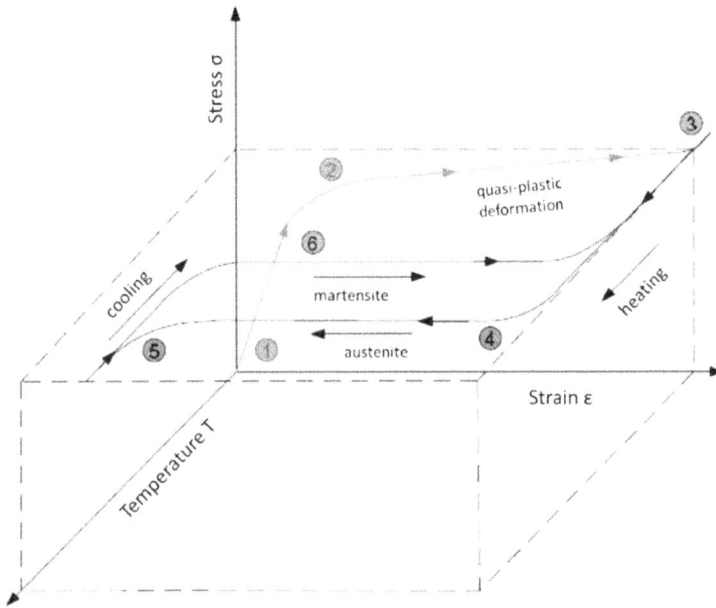

Figure 3.
Schematic of the extrinsic two-way effect in a σ(ε, T) diagram according to Ref. [18].

temperature. To overcome this problem, various approaches are being investigated in current research, such as the use of a second, antagonistically arranged actuator element. Due to the flat course in the pseudoplastic region, at least theoretically the possibility of an energy-free holding of the position arises.

When operating an SMA actuator, it is possible to move to any intermediate position in its working range. This can only be done by a controlled partial phase transformation. This means that in a defined temperature interval, specific temperatures are approached and kept constant.

The generally high-operating material stresses make SMs the actuator mechanism with the highest energy density of any known technology [19–21]. This means that very high forces can be generated with only a small amount of material, which in turn can lead to very compact and lightweight actuator systems. Typical energy densities are in the order of 10^7 J/m^3 [22]. The strokes that can be generated are also comparatively high with typically 3–5% for NiTi actuator materials [23–25]. The limiting design parameter for SMA actuators is usually the switching frequency, which is determined by the thermal cooling behavior of the SMA element, and thus rarely exceeds an order of 10 Hz. In addition, the temperature range in the application must be considered, as today's commercial alloys have phase transition temperatures in the 90°C range. At ambient temperatures in this range, it therefore becomes difficult to further utilize the actuator properties.

In most SMA actuator system developments today, the focus is on the design of tensile loaded SMA wire actuators. Wire actuators are usually activated by means of electrical current through the inherent resistance of the shape memory material. In addition to wire actuators, there are also SMA springs, tubes, bundles and other designs, often specifically developed for applications [26–29].

3.1 SMA actuator system design

As described above, the realization of SMA actuators requires loads or preload elements that, if possible, return the SMA actuator to its fully de-twinned initial state upon cooling. Using a wire actuator as an example, three common actuator systems are presented: constant load, linear spring load, and antagonistic actuators.

In the simplest case, the actuator is reset by a constant load (e.g., a mass). **Figure 4** (upper part) shows the mechanical principle schematically [30]. The cold SMA wire is stretched in the martensitic state by the mass up to the corresponding mechanical stress level. Heating of the wire leads to phase transformation and the associated contraction of the wire. In the stress-strain diagram, the constant load now intersects the red austenite curve. Cooling again leads to elongation of the SMA wire.

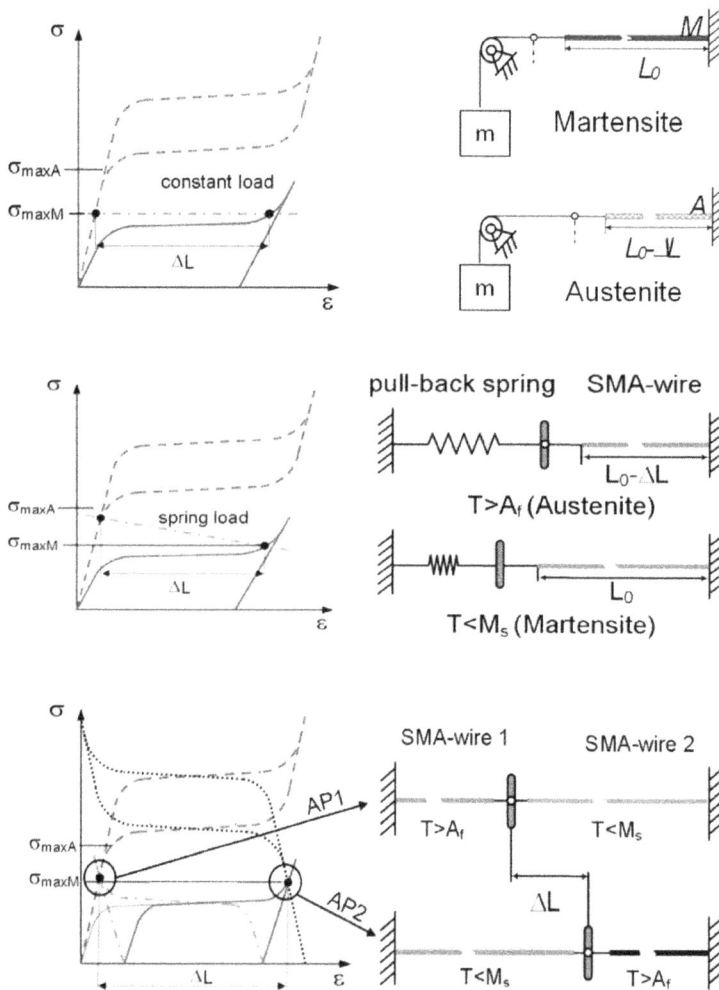

Figure 4.
SMA wire actuator systems with different preload and resetting principles: Constant load (upper part), linear spring load (center part), and antagonistic SMA wires (lower part) [30].

In case of a linear spring used as a restoring element (**Figure 4**, center part), the mechanical stress increases when the SMA wire is activated, and the stroke is also slightly lower compared to the constant load system due to the Young's modulus in the austenite.

Compared to the wire-spring configuration, antagonistic SMA wires only use up the necessary material stress to remain at a certain operating point (**Figure 4**, lower part). In addition, energy-free (no electrical current flow) holding of positions up to a certain stress level is possible in this system.

3.2 Energy-efficient SMA actuator systems

With a focus on energy-efficiency, a possible approach is to couple SMA wires with bistable elements (springs) [31–34]. This results in a switching bistable actuator (**Figure 5**), which is held in its two end positions solely by the stiffness of the bistable element, leading to better energy-efficiency compared to conventional SMA actuators. In addition, the mechanical design allows more space-efficient solutions to be realized and even actuator strokes to be significantly increased.

Another possibility to increase the performance of SMA actuators and also to operate them more energy-efficiently is *via* control strategy. The electrical activation of SMA elements usually takes place in the low-voltage range. A certain amount of energy, dependent on the total SMA mass, is required for heating to the phase transition temperature and then a specific amount of latent heat for the phase transformation itself. Ideally, SMA actuators are "adiabatically" activated, meaning that no energy is lost to the environment during the activation or heating interval. Electrical power therefore plays a significant role. With targeted energy quantity delivery, SMA actuators can therefore be operated at high voltages and power levels. This type of activation strategy is studied and explained in detail in Refs. [35–38]. In these investigations, switching times in the lower millisecond range were demonstrated at electrical voltages in the 100 V range, for example, as well as energy savings of up to 80% compared to conventional electrical activation.

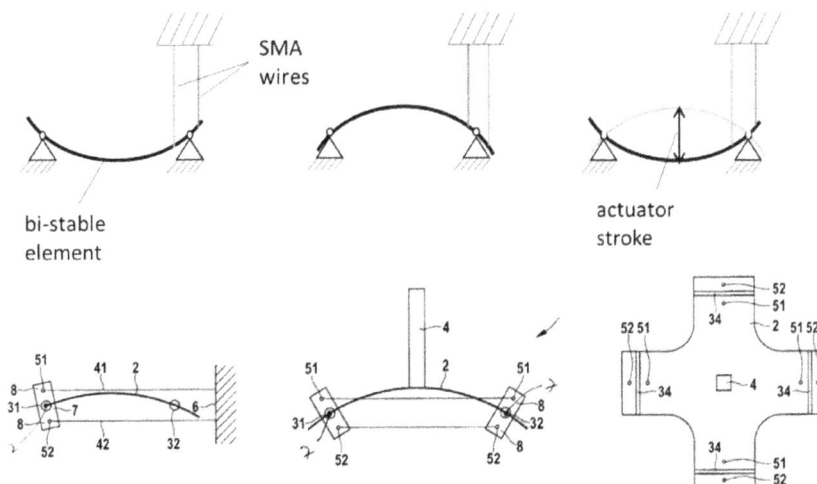

Figure 5.
Various exemplary configurations of bistable springs and SMA wires [31–34].

4. SMA actuator systems and applications

In the following, an overview of realized SMA actuators in prototype stage and commercial applications is given. A distinction is made between switching systems, closed-loop controlled systems, and energy-autonomous systems. In principle, a large number of SMA actuators are known today. The focus of this section is on systems, which are preferably close to series production or available on the market and which prove the potential of SMA due to their successful assertion on the market. A comprehensive overview of SMA actuator systems is presented in Ref. [9].

4.1 Switching systems

Switching systems use the shape memory effect to realize two defined switching states. In most cases, such systems are designed as wire-spring arrangements, with the SMA elements activated either by externally applied thermal energy or by electrical current. Although switching actuators have a low potential for possible functional integration into systems and can only be adapted to the individual requirements of an application to a limited extent compared to controlled or energy-autonomous systems, there are nevertheless immense advantages to be gained from their use. In addition to the reduction of installation space, mass, and complexity, the silent operation can also be an important trait. Established applications for switching actuators can be found in valve technology, among others. This includes thermally activated valves in cooling circuits (automotive) or in coffee machines, water valves with scald protection [39, 40], but also electrically activated valves for pneumatic applications. Particularly noteworthy in this context are the SMA valve for thermal management from Ingpuls GmbH, and the 3/3-way valve for controlling the massage function in passenger car seats developed by Actuator Solutions GmbH—a joint venture of the companies Alfmeier Präzision AG and SAES Getters (**Figure 6**) [41]. Ingpuls generally acts as a material, component, and system manufacturer and has already equipped more than 5 million vehicles with actuators in thermal management. In addition to the automotive industry, the company also serves other sectors such as aerospace, medical technology, and household appliances.

Actuator Solutions' electrically activated valves have a total mass of less than 20 g. In these valves, spring-loaded SMA wires in a V arrangement are energized to switch two plungers with a stroke of 0.4 mm at a force of 0.8 N between two end positions completely silently. The small cross section of the FGL wires used, 76–100 μm, permits low activation currents and thus comparatively low energy consumption. The

Figure 6.
Pneumatic SMA valve from actuator solution GmbH [41, 42].

market launch took place in 2005 with an annual volume of 150,000 units. Since then, the market volume increased to around 15 million units per year by 2015. Production of the valves is fully automated.

An application example for a bistable switching system with SMA wires is a bistable vacuum suction cup for material handling [43]. The actuator system consists of a bistable spring in cross form and two antagonistic bundles of SMA wires, which are attached to the rotationally mounted spring element *via* small levers (**Figure 7**). Alternating activation of the wire bundles moves a flexible membrane from a flat to a deformed state and back again. As in a suction cup, this can be used to grip structurally rigid, flat components.

4.2 Controlled systems

In order to position accurately with SMA actuators, closed-loop control is necessary. Compared to purely switching actuators, closed-loop systems are significantly more complex, since they must have additional sensors and a control device. So far, two systems from the field of optics have established themselves on the market in which closed-loop controlled SMA actuators are used. In many smartphone cameras, two systems are integrated with which the lens is finely positioned. On the one hand, AF (Auto Focus) enables optimally focused images through translational positioning. Secondly, OIS (Optical Image Stabilization) compensates for low-frequency vibrations caused by hand movements through tilting the lens in two degrees of freedom. This achieves better image quality, especially in low-light conditions. While the AF function can be realized conventionally by comparatively simple voice coil drives, the forces achievable by the coils are too low for the implementation of OIS systems. Due to their extraordinary energy density, SMAs are predestined to implement such functions in the smallest installation space with significantly lower weight. **Figure 8** upper part shows an SMA-based AF system. A spring-loaded wire in a V arrangement, which is crimped to the housing, is heated with electric current (red) and then

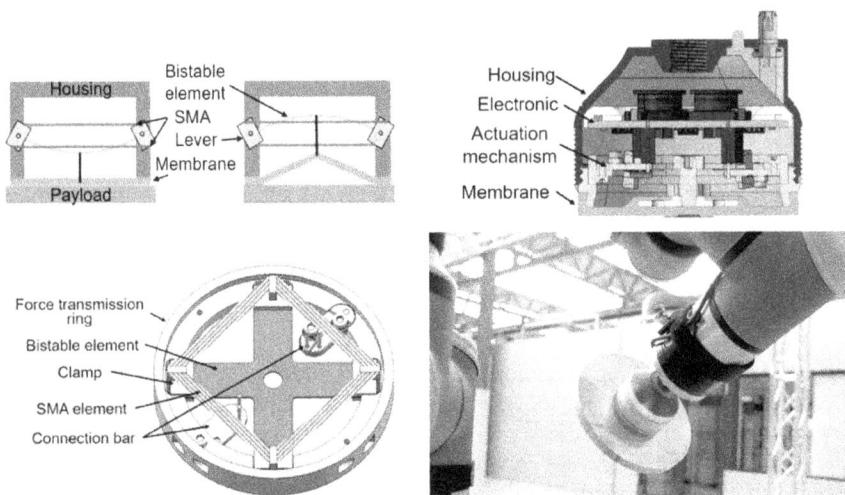

Figure 7.
Schematic operation and design and assembly of the SMA-based vacuum gripper [43].

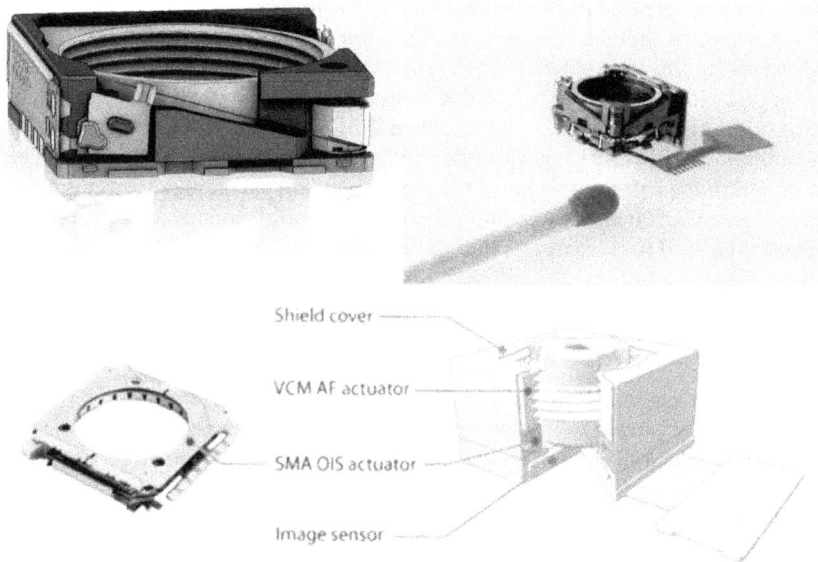

Figure 8.
SMA-based AF (upper part) and OIS (lower part) actuators in smart phone camera systems [44].

contracts, lifting the spring-loaded lens carrier. The position is detected indirectly by measuring and evaluating the electrical resistance of the wire.

In OIS systems, two actuators in a V arrangement are required to realize the two tilting degrees of freedom. The comparatively small stroke or tilt angle thus allows the very flat design shown in **Figure 8** lower part. This makes it possible to arrange the SMA-based OIS under conventional AF systems without increasing the installation space. The control and position measurement of the two actuators can also be performed here indirectly *via* the measurement of the electrical resistance of the SMA wires.

In the industrial field of handling technology, an adaptive end-effector has been developed in combination with the vacuum grippers already mentioned [45] All four gripper arms can be positioned independently of each other in a position-controlled manner. Thereby, each gripper arm has the possibility to rotate up to 90° in plane, as well as to tilt up to 30° out of plane (**Figure 9**, upper part). These degrees of freedom enable the handling of various component geometries in production without having to replace the specific end-effectors required for this purpose. SMA wires are attractive in this case primarily because of their light weight, as additional electrical or pneumatic actuators would introduce a non-economic increase in inertial mass.

Another current research field in which SMA actuators are gaining relevance is continuum robotics. Inspired by elephant trunks, tentacles, or snakes in nature, robotic structures are being developed without discrete mechanical joints [48, 49]. SMA wires are suitable as manipulators for robotic structures, for example, in the medical field in endoscopes, guide wires or catheters. **Figure 9** lower part shows three different designs of such continuum robots. Lower part right shows an endoscopic camera system for the inspection of complex components. By actively controlling the tip of the endoscope on which the camera modules sit, aircraft turbines or turbochargers, for example, can be inspected from the inside for material defects or damage.

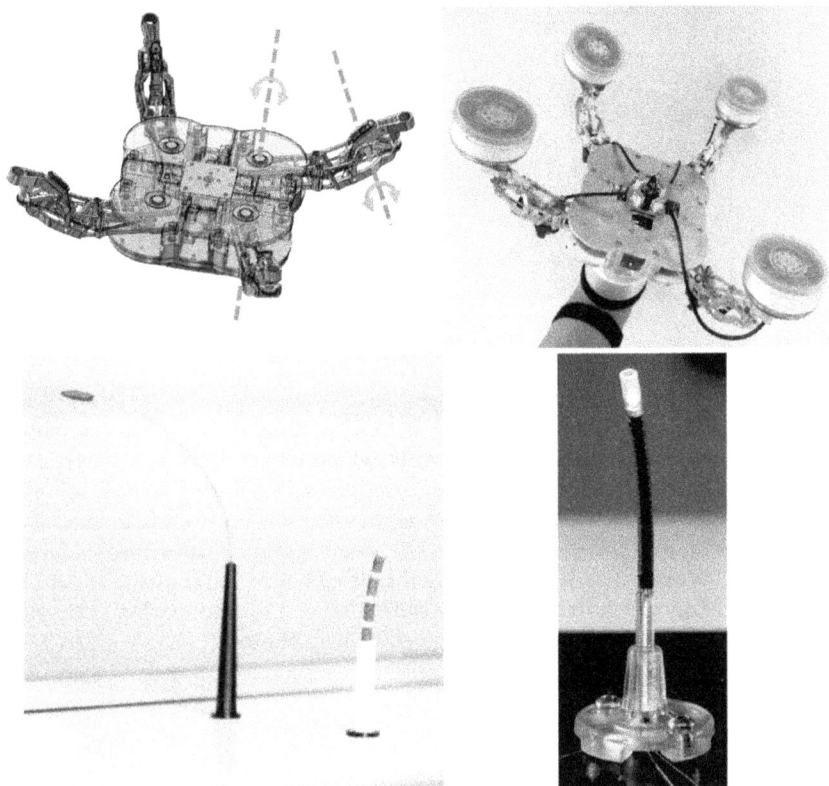

Figure 9.
SMA-based end-effector (upper part) [45] and SMA soft continuum robotic structures (lower part) [46, 47] (photo: Oliver Dietze).

4.3 Energy-autonomous systems

While switching and controlled SMA actuators always require an external energy supply and control or regulation components, energy-autonomous SMA actuators are characterized by the fact that they are activated by the surrounding medium. Actuator, sensor, and control functions are realized solely by the SMA material. This allows a very high degree of simplicity and functional integration, but requires multidisciplinary knowledge in material science, thermodynamics, design, and control engineering to develop such systems. Several applications are known from the state-of-the-art technique. These range from thermostatic valves and valves for compensation of viscosity differences to mechanical engineering components for compensation of thermally induced bias losses [50].

Another application from the aerospace sector, which is characterized by particularly high functional integration, is described in Refs. [51, 52]. This system, which can be found in the literature as a VGC (variable geometry chevron), solves the conflict between low noise emissions during takeoff and landing and high efficiency at cruising speed. At the outlet of the thrust nozzle of modern jet engines, there are often sawtooth-shaped guide elements, the so-called chevrons. These serve to improve mixing of the air layers between the hot and fast combustion gases from the thrust

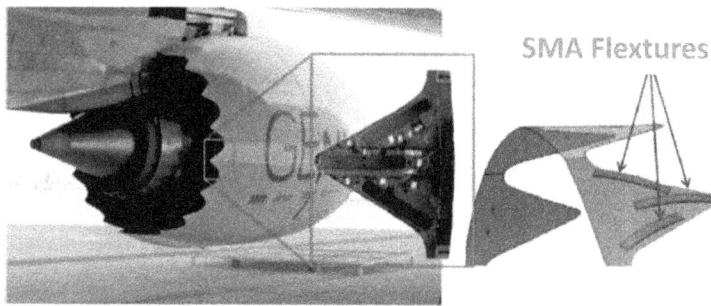

Figure 10.
Variable geometry chevron with integrated SMA actuators [53].

nozzle and the cold and slower jacket flow. In VGC systems, SMA bending actuators are embedded in these composite components, which are activated by the exhaust gas temperature (**Figure 10**). The high temperatures prevailing during takeoff initially deform the conductive elements inward, thus ensuring low noise emissions from the engine. As altitude and speed increase and the ambient temperature drops as a result, they deform back outward, thus increasing the efficiency of the engine. This application is a structurally integrated actuator system, that is, a system in which the solid-state properties of the SMA material are specifically exploited.

5. SMA actuator materials

Technically relevant shape memory alloys today are NiTi, NiTi(−X) (addition of further elements X), CuZnAl, and CuAlNi. NiTi and NiTi(−X) alloys offer the most balanced property profile. With respect to mechanical properties, CuZnAl and CuAlNi alloys exhibit some significant disadvantages. In particular, the smaller effect size (of shape memory effect) and the limited cycle stability should be mentioned. For this reason, NiTi alloys have become widely accepted today and are still the only alloys that are commercially available to any significant extent.

In the field of NiTi or NiTi(−X) alloys, the binary NiTi alloys with approximately stoichiometric composition have the greatest technical relevance [54]. The position of the phase transformation temperature is adjusted by the concentration ratio of the alloying partners Ni and Ti. It should be noted that even small deviations in the composition strongly influence the transformation temperature level. For example, a shift in the composition by approx. 0.1 atomic percent already means a shift in transformation temperature by approx. 10 K. Accordingly, the manufacturing process must be very precise.

Alloy-related disadvantages of the binary NiTi alloy can be compensated by adding further components (NiTi(−X)). The focus of research here is on shifting the transformation temperatures to higher temperatures (X: platinum, palladium, hafnium, zirconium), lower thermal hysteresis (X: copper), and the most linear possible relationship between martensite content and electrical resistance (X: copper) for self-sensing applications [55]. So far, only the alloy system NiTiCu has gained some importance. Compared to binary NiTi alloys, however, the availability of suitable semi-finished products is severely limited. The basic properties of different alloy systems are summarized in **Table 1**.

Characteristic	NiTi/NiTi(−X)	CuZnAl	CuAlNi
Density [g/cm^3]	6.4...6.5	7.6...7.65	7.1...7.15
Specific electrical resistance [10^{-6} Ωm]	0.5...1.1	0.07...0.12	0.1...0.14
Thermal conductivity at 20 °C [W/(m•K)]	10...18	120	75
Thermal expansion coefficient [10^{-6} K^{-1}]	6.6...10	17	17
Specific heat capacity [J/(Kg•K)]	490	390	440
Transformation enthalpy [J/kg]	28,000	7000	9000
Young's modulus [GPa] (Austenite)	95	70...100	80...100
Fatigue stress [MPa]	250	75	100
Transformation temperatures [°C]	−50...100	−100...100	−100...170
Thermal hysteresis (A_s-M_f) [K]	20...40	10...20	20...25
Spread (A_f-A_s) [K]	30	10...20	20...30
Max. one-way effect [%]	8	4	5
Max. two-way effect [%]	4	1	1.2
Typical fatigue cycles	1,000,000	100,000	10,000

Table 1.
Properties of technically important shape memory materials ([18, 55, 56]).

6. Conclusions

Thermal shape memory alloys have gained significant technological importance in the recent decades. The outstanding fatigue properties combined with resistance to corrosion and biocompatibility led to the breakthrough in the use of superelastic SMA in medical technology. Stents made of pseudoelastic nickel-titanium are used to dilate blood vessels. The rhythmic pulsation of the blood circulation causes dynamic loading of the stent with large strain amplitudes. It is evident that the stent material must be highly fatigue resistant to endure the enormous numbers of cycles during continuous use in the human body. In this context, the pseudoelasticity allows repeatable strains up to 8%. Depending on the thermal and mechanical load, cycle numbers of 10^7 and higher can be achieved today.

Due to the medical demand for nickel-titanium, this material is nowadays reproducibly available in large quantities worldwide at favorable prices. Due to their high energy density, SMAs are also becoming increasingly attractive for actuator applications. The potential of miniaturization, weight and energy savings, and the combination of actuator and sensing properties will also lead to more and more commercial applications for SMA actuators in the future.

For commercial systems with interesting economic quantities and yearly produced piece numbers, the logical focus is on SMA wire-based actuator systems, as the semi-finished product (the SMA wire) is available in good and reproducible quality from multiple world-wide sources in sufficient quantities. The main challenges for SMA wire-based actuator systems are the realization of efficient, fatigue-resistant systems with sufficient cycle-times in operation temperatures. While material development is focusing on aspects like improved fatigue behavior through microstructure and surface finish as well as increasing phase transformation temperatures for the realization of a wider range of applications, system engineering should put focus on the following aspects: Realization of higher actuation frequencies by implementation of bundled micro-diameter wires, increase of actuator stroke by specific kinematic designs, and smart system implementation to decrease total energy consumption.

The exploitation of the materials self-sensing ability will further enable the development of smarter drive solutions and add substantial benefits in competition with conventional actuator solutions. The handling of the hysteretic resistance feedback of SMA wires in applications as well as novel control algorithms might be the field of research with the biggest importance today.

Acknowledgements

The authors would like to thank André Bucht, Kenny Pagel, and Thomas Mäder for the pleasant collaboration on an SMA book project in the past, which served as a general guideline for this book chapter.

Author details

Paul Motzki[1]* and Gianluca Rizzello[2]

1 Center for Mechatronics and Automation Technology (ZeMA), Saarland University, Saarbruecken, Germany

2 Saarland University, Saarbruecken, Germany

*Address all correspondence to: paul.motzki@uni-saarland.de

IntechOpen

References

[1] Motzki P, Bucht A, Pagel K, Mäder T, Seelecke S. Thermische Formgedächtnislegierungen. In: Böse H, editor. Smart Materials - Eigenschaften und Einsatzpotenziale. 1st ed. Würzburg: Vogel Communications Group; 2023. pp. 89-118 [Online]. Available from: https://vogel-fachbuch.de/detail/index/sArticle/1086

[2] Reece PL. Progress in Smart Materials and Structures. New York: Nova Science Publishers, Inc.; 2007

[3] Chang LC, Read TA. Transducers. Amer. Inst. Min. (Metall.) Engrs. 1951;**191**:47

[4] Chang LC. Atomic displacements and crystallographic mechanism in Diffusionless transformation of gold-cadmium single crystals containing 475 atomic percent cadmium. Acta Crystallographica. 1951;**4**:320-324

[5] W. J. Buehler, J. V. Gilfrich, and R. C. Wiley, Effect of low-temperature phase changes on the mechanical properties of alloys near composition TiNi, Journal of Applied Physics, vol. 34, no. 5, pp. 1475-1477, May 1963, doi: 10.1063/1.1729603.

[6] Janocha H. Adaptronics and Smart Structures, 2. Berlin Heidelberg: Springer Verlag; 2007. DOI: 10.1017/CBO9781107415324.004

[7] Janocha H, Bonertz T, Pappert G. Unkonventionelle Aktoren : eine Einführung. München: Oldenbourg Wissenschaftsverlag; 2013

[8] Motzki P. Advanced Design and Control Concepts for Actuators Based on Shape Memory Alloy Wires. Dissertation, Saarland University. 2018. DOI: 10.22028/D291-27354

[9] Mohd Jani J, Leary M, Subic A, Gibson MA. A review of shape memory alloy research, applications and opportunities. Materials and Design. 2014;**56**:1078-1113. DOI: 10.1016/j.matdes.2013.11.084

[10] Van Humbeeck J, Chandrasekaran M, Delaey L. Shape memory alloys: Materials in action. Endeavour. 1991;**15**(4):148-154. DOI: 10.1016/0160-9327(91)90119-V

[11] Lagoudas DC. Shape Memory Alloys. Vol. 1. Boston, MA: Springer US; 2008. DOI: 10.1007/978-0-387-47685-8

[12] Funakubo H. Shape Memory Alloys. Vol. 1, no. D. Amsterdam: Gordon and Breach Science Publ; 1987. DOI: 10.1007/978-0-387-47685-8

[13] Just E. Entwicklung eines Formgedächtnis-Mikrogreifers, Dissertation, Universität Karlsruhe. 2001

[14] Duerig TW, Melton KN, Stöckel D, Wayman CM. Engineering Aspects of Shape Memory Alloys. London: Butterworth-Heinemann; 1990

[15] Lagoudas DC. Shape Memory Alloys: Modeling and Engineering Applications. New York: Springer; 2008

[16] Maaß B. Strukturbildungsprozesse und funktionelle Eigenschaften bei der Herstellung pseudoelastischer Ni-Ti-Cu-(X)- Formgedächtnislegierungen. Dissertation, Ruhr-Universität Bochum. 2012

[17] Otsuka K, Wayman CM. Shape Memory Materials. Cambridge: Cambridge University Press; 1998

[18] Langbein S, Czechowicz A. Konstruktionspraxis

Formgedächtnistechnik. Vol. 1. Wiesbaden: Springer Fachmedien; 2013. DOI: 10.1017/CBO9781107415324.004

[19] Crews JH. Development of a shape memory alloy actuated robotic catheter for endocardial ablation: Modeling, design optimization, and control. 2011 [Accessed: Oct. 22, 2015] [Online]. Available from: http://gradworks.umi.com/34/63/3463761.html

[20] Zimmerman E, Muntean V, Melz T, Seipel B, Koch T. Novel pre-crash-actuator-system based on SMA for enhancing side impact safety. In: Meyer G, Valldorf J, Gessner W, editors. Advanced Microsystems for Automotive Applications. VDI-Buch. Berlin, Heidelberg: Springer; 2009. DOI: 10.1007/978-3-642-00745-3_4

[21] Kohl M, Krevet B, Just E. SMA microgripper system. Sensors and Actuators A: Physical. 2002;**97-98**:646-652. DOI: 10.1016/S0924-4247(01)00803-2

[22] Janocha H. Unkonventionelle Aktoren. München: Oldenbourg Wissenschaftsverlag; 2013. DOI: 10.1524/9783486756920/HTML

[23] SAES Getters. SmartFlex Brochure. 2017. Available from: https://www.saesgetters.com/sites/default/files/SmartFlexBrochure_2.pdf [accessed Nov. 15, 2020]

[24] Dynalloy Inc. Technical Characteristics of Flexinol Actuator Wires. Datasheet; 2019. Available from: http://www.dynalloy.com/pdfs/TCF1140.pdf [accessed May 15, 2021]

[25] Ingpuls GmbH. Ingpuls GmbH – Produktportfolio. 2020. https://ingpuls.de/produkte-leistungen/komponenten.html [accessed May 15, 2021]

[26] Britz R, Motzki P. Analysis and evaluation of bundled SMA actuator

wires. Sensors and Actuators A: Physical. 2022;**333**:113233. DOI: 10.1016/j.sna.2021.113233

[27] Molitor P, Britz R, Motzki P. High-power shape memory alloy catapult actuator for high-speed and high-force applications. IEEE Access. 2022;**10**:92373-92380. DOI: 10.1109/ACCESS.2022.3202210

[28] Britz R, Rizzello G, Motzki P. High-speed antagonistic shape memory actuator for high ambient temperatures. Advanced Engineering Materials. 2022;**24**:2200205. DOI: 10.1002/adem.202200205

[29] Britz R, Motzki P, Seelecke S. Thermal actuator arrangement having improved reset time. WO2021052933A1, US17761385. 2019

[30] VDE/VDI-Gesellschaft Mikoelektronik Mikro- und Feinwerktechnik (GMM). Innovative Kleinantriebe: Tagung Mainz. In: Innovative Kleinantriebe: Tagung Mainz. Mainz: VDI Verlag; 1996

[31] Motzki P, Seelecke S. Bi-stable SMA actuator. In: Borgmann H, editor. Actuator 16 - 15th International Conference on New Actuators. Bremen, Germany: MESSE BREMEN; 2016. pp. 317-320. DOI: 10.13140/RG.2.2.12065.20325

[32] Motzki P, Seelecke S. Bistabile Aktorvorrichtung mit einem Formgedächtniselement DE 10 2016 108 627 A1. 2016 [Online]. Available from: https://depatisnet.dpma.de/DepatisNet/depatisnet?action=bibdat&docid=DE102016108627A1

[33] Motzki P, Seelecke S. Bistable actuator device having a shape memory element. WO 2017/194591 A1. 2016 [Online]. Available from: https://depatisnet.dpma.de/DepatisNet/depatisnet?action=bibdat&docid=WO0020171 94591A1

[34] Motzki P, Seelecke S. Bistable actuator device having a shape memory element. US 2019/0203701 A1. 2019

[35] Vollach S, Shilo D. The mechanical response of shape memory alloys under a rapid heating pulse. Experimental Mechanics. 2010;**50**(6):803-811. DOI: 10.1007/s11340-009-9320-z

[36] Dana A, Vollach S, Shilo D. Use the force: Review of high-rate actuation of shape memory alloys. Actuators. 2021;**10**(7):140. DOI: 10.3390/ACT10070140

[37] Vollach S, Shilo D, Shlagman H. Mechanical response of shape memory alloys under a rapid heating pulse - part II. Experimental Mechanics. 2016;**56**(8):1465-1475. DOI: 10.1007/s11340-016-0172-z

[38] Motzki P, Gorges T, Kappel M, Schmidt M, Rizzello G, Seelecke S. High-speed and high-efficiency shape memory alloy actuation. Smart Materials and Structures. 2018;**27**(7):075047. DOI: 10.1088/1361-665X/aac9e1

[39] Ingpuls GmbH. Produkte und Leistungen. [Online]. Available from: https://ingpuls.de/produkte-leistungen/

[40] Duerig TW. Applications of shape memory. In: Materials Science Forum. Vols. 56-58. Trans Tech Publications, Ltd.; 1991. pp. 679-691. DOI: 10.4028/www.scientific.net/msf.56-58.679

[41] Actuator Solutions GmbH. Actuator solutions SMA products. 2018. Available from: http://www.actuatorsolutions.de/products/ [accessed Nov. 15, 2020]

[42] Actuator solutions GmbH. Pictures sent to me by Actuator solutions GmbH

[43] Motzki P, Seelecke S. Industrial applications for shape memory alloys. In: Olabi A-G, editor. Encyclopedia of Smart Materials. Elsevier; 2022. pp. 254-266. DOI: 10.1016/B978-0-12-803581-8.11723-0. Available from: https://www.sciencedirect.com/science/article/abs/pii/B9780128035818117230

[44] Cambridge Mechatronics Ltd. CML OIS actuator. [Accessed: Apr. 13, 2020]. [Online]. Available from: https://www.cambridgemechatronics.com/en/cml-technology/actuators/

[45] Motzki P, Khelfa F, Zimmer L, Schmidt M, Seelecke S. Design and validation of a reconfigurable robotic end-effector based on shape memory alloys. IEEE/ASME Transactions on Mechatronics. 2019;**24**(1):293-303

[46] Mandolino MA, Goergen Y, Motzki P, Rizzello G. Design and characterization of a fully integrated continuum robot actuated by shape memory alloy wires. In: 2022 IEEE 17th International Conference on Advanced Motion Control (AMC), Padova, Italy. 2022. pp. 6-11. DOI: 10.1109/AMC51637.2022.9729267

[47] Goergen Y, Rizzello G, Seelecke S, Motzki P. Modular design of an SMA driven continuum robot. In: Proceedings of the ASME 2020 Conference on Smart Materials, Adaptive Structures and Intelligent Systems. ASME 2020 Conference on Smart Materials, Adaptive Structures and Intelligent Systems. Virtual, Online. September 15, 2020. V001T04A007. ASME. DOI: 10.1115/SMASIS2020-2213

[48] McMahan W, Jones BA, Walker ID. Design and implementation of a multi-section continuum robot: Air-Octor

[49] Cowan LS, Walker ID. The importance of continuous and discrete elements in continuum robots. International Journal of Advanced

Robotic Systems. 2013;**10**(3).
DOI: 10.5772/55270

[50] de Navarro y Sosa I, Bucht A,
Junker T, et al. Novel compensation of
axial thermal expansion in ball screw
drives. Production Engineering: Research
and Development. 2014;**8**:397-406.
DOI: 10.1007/s11740-014-0528-0

[51] Calkins FT, Mabe JH. Shape
memory alloy based morphing
Aerostructures. Journal of Mechanical
Design. 2010;**132**(11):111012.
DOI: 10.1115/1.4001119

[52] Mabe JH, Calkins FT,
Alkislar MB. Variable area jet nozzle
using shape memory alloy actuators in an
antagonistic design. In: Proceedings of
SPIE 6930, Industrial and Commercial
Applications of Smart Structures
Technologies. 69300T. 22 April 2008.
DOI: 10.1117/12.776816

[53] Oehler SD, Hartl DJ, Lopez R,
Malak RJ, Lagoudas DC. Design
optimization and uncertainty
analysis of SMA morphing
structures. Smart Materials and
Structures. 2012;**21**(9):094016.
DOI: 10.1088/0964-1726/21/9/094016

[54] Mertmann M. NiTi-
Formgedächtnislegierungen für Aktoren
der Greifertechnik. Fortschrittsbericht
VDI Reihe 5 Nr. 469, Düsseldorf. 1997.

[55] C. Lexcellent. Shape-
memory Alloys Handbook. 2013.
DOI: 10.1002/9781118577776

[56] Gümpel P. Formgedächtnislegierungen
in Maschinenbau, Medizintechnik und
Aktuatorik. Renningen: Expert-Verlag;
2004

Section 2

Design and Development of Shape Memory Alloys

Chapter 3

Design and Development of B2 $Ti_{50}Pd_{50-x}M_x$ (M = Os, Ru, Co) as Potential High-Temperature Shape Memory Alloys

Ramogohlo Diale, Phuti Ngoepe and Hasani Chauke

Abstract

In this chapter, the stability and phase transformation of B2 $Ti_{50}Pd_{50-x}M_x$ (M = Os, Ru, Co) alloys are investigated using density functional theory. TiPd alloy can be suitable for high-temperature shape memory applications due to its martensitic transformation capability from B2 to B19 at 823 K. It was reported that the binary $Ti_{50}Pd_{50}$ alloy is mechanically unstable at 0 K. A partial substitution of Pd with Os, Ru, or Co is investigated to determine which alloy will have the best properties. The heat of formation, density of states, and mechanical properties were determined to check the stability. The heat of formation was found to decrease with an increase in Ru and Os concentrations (condition of stability), consistent with the density of states trend. This is in contrast to Co addition, which indicates that the thermodynamic stability is not enhanced (heat of formation increases). It was also found that an increase in Os, Ru, and Co content stabilizes the $Ti_{50}Pd_{50}$ with a positive elastic shear modulus ($C'>0$) above 18.25, 20, and 31 at.%, respectively. This book chapter will provide valuable insights to guide experiments in the design and development of alloys.

Keywords: B2 $Ti_{50}Pd_{50-x}M_x$, shape memory alloys, DFT, stability, phase transformation

1. Introduction

Shape-memory alloys (SMAs) are known as smart materials which exhibit unusual elastic and mechanical behaviors, such as shape memory effect and superelasticity [1]. These materials are capable of restoring their shape after being excessively deformed at low or high temperatures. SMAs undergo a reversible martensitic phase transformation upon the influence of temperature or stress field, giving rise to the shape memory effect and superelasticity [1]. Recently, SMAs are being developed to suit many applications in many fields, especially for engineering properties. The most commonly known SMAs include nickel-titanium (NiTi), nickel-titanium copper (NiTiCu)), and many other metallic alloy systems where the application temperatures do not exceed 373 K [2]. There is an increasing demand for high-temperature shape memory alloys (HTSMAs) in actuators for automobile, pipe couplings and aircraft engines, and other applications. Recently, new alloy systems which can work above that temperature have been investigated such as TiPt, TiAu, and TiPd binary systems as possible new HTSMA materials [3]. As their transformation temperatures are above 800 K and have martensitic transformations from B2 in the austenite phase to B19 in the martensite phase [4].

$Ti_{50}Pd_{50}$ systems are considered as one of the potential high-temperature shape memory alloys (HTSMA's) due to their high martensitic transformation temperature [5, 6]. The $Ti_{50}Pd_{50}$ has two stable phases—the high-temperature phase called austenite and the low-temperature phase called martensite [7]. The $Ti_{50}Pd_{50}$ has a simple cubic CsCl-type structure (cP2, B2) at high temperatures [8], and an orthorhombic AuCd-type structure (oP4, B19) at ambient temperatures [9, 10]. Besides being lightweight and oxidation resistant, these alloys are also ductile at 823 K [11, 12]. Previously, B2 $Ti_{50}Pd_{50}$ was reported to be mechanically unstable due to negative elastic shear modulus [4, 11, 13, 14]. Furthermore, it was also reported that the strength of these alloys drop above 823 K. As a result, there is a need to establish ternary alloys to improve the properties of the binary alloy $Ti_{50}Pd_{50}$ that can be used for actuators and the aeronautics industry [15, 16]. The collapse is due to a possible phase transformation from body-centered cubic (bcc) to other tetragonal and orthorhombic phases, such as $L1_0$ and B19, similar to those reported in TiPt alloy [17]. Nonetheless, their transformation behavior has not been ascertained explicitly.

In order to enhance the transformation temperature and performance of the $Ti_{50}Pd_{50}$ SMA's ternary alloying has been suggested. The elements such as Au, Ni, Ru, Rh, Ir, Pt, Zn, Rc, Tc, Os, and Co were reported as the best site preference for both Ti and Pd substitution with less than 50 atomic percentage (at. %), while Ag and Cd prefer the Ti substitution site in B2 $Ti_{50}Pd_{50}$ structure [18]. A high work output was shown and good workability was demonstrated by the addition of the third elements such as Ru, Ir, Pt, Co, and Ni. [19]. Previously, the addition of Ni to $Ti_{50}Pd_{50}$ has shown improvement in shape memory characteristics for $Ti_{50}Pd_{30}Ni_{20}$ composition [11]. Furthermore, another study was conducted on the ternary alloying of TiPd with Ru addition using density functional theory (DFT)) [20]. The findings showed that the mechanical stability of $Ti_{50}Pd_{50}$ is enhanced above 25 at. % Ru. A cluster expansion approach was also used in our other study to investigate the phase stability of $TiPd_{1-x}Ru_x$ and $Ti_{1-x}PdRu_x$ shape memory alloys [21]. As a result of the research, it was found that Ru prefers Ti-site compared to Pd-site, which showed better thermodynamic stability. In a previous study [22], DFTB+ code was used to develop a set of potential parameters of $Ti_{50}Pd_{50-x}Ru_x$ shape memory alloys. A set of potential parameters were developed in order to investigate the stability and transformation temperature of investigated alloys. It was found that the addition of 6.25, 18.75, and 25 at. % Ru on $Ti_{50}Pd_{50}$ reduce the transformation temperature from B19 to B2 phase.

Cobalt and other PGMs elements have been in use for the production of various components such as vanes or combustion chambers in gas turbines for their exceptional heat-resistant properties [23]. In previous studies, it was discovered that Ru increased the mechanical stability of high-temperature single-crystal superalloys and titanium-based alloys for use in jet engines [24, 25]. Due to their better temperature capabilities and reduced creep rate, these alloys will boost aircraft efficiency and increase the durability of the actuator [24]. Osmium is regularly utilized as an alloying element with other PGMs in order to enhance electricity and stiffness in medical devices and other applications [26]. The ternary alloying with Os, Ru, and Co may improve the stability and the transformation temperature to above 1000 K [27].

In this chapter, the ternary alloying of Ti$_{50}$Pd$_{50}$ with Os, Ru, and Co has been performed using the density functional theory approach to investigate the thermodynamic, electronic, and mechanical stability. Furthermore, the effect of Os, Ru, and Co on the ductility/brittleness has been deduced from the anisotropy ratio, which confirms the strength of the systems. The findings will provide very valuable information and practical guidance on the development of these alloys in the future.

2. Computational methodology

2.1 Density functional theory (DFT)

Density functional theory (DFT) is a quantum mechanical theory used in physics and chemistry to study the electronic structure and ground state properties of many-body systems, particularly molecules, atoms, and condensed phase. DFT was first formulated by Hohenberg and Kohn in 1964 [28], then secondly developed by Kohn and Sham in 1965 [29]. Using DFT, independent particle methods have been developed that take into account particle correlations and interactions. The ground state properties of a many-electron system are determined by an electron density that is dependent on the three spatial coordinates as follows:

$$E = E\left[\rho\left(\vec{r}\right)\right], \tag{1}$$

where E is the total energy and ρ is the density.

Kohn and Sham further derived different sets of differential equations which enable the calculation of ground state density $\rho_0\left(\vec{r}\right)$ to be found. The ground state energy of the electronic structure is calculated from the following equation:

$$E\left[\rho\left(\vec{r}\right)\right] = T_s\left[\rho\left(\vec{r}\right)\right] + \frac{1}{2}\iint \frac{\rho\left(\vec{r}\right)\rho\left(\vec{r}'\right)}{|\vec{r} - \vec{r}'|} d\vec{r}d\vec{r}' + E_{XC}\left[\rho\left(\vec{r}\right)\right] + \int \rho\left(\vec{r}\right)V_{ext}\left(\vec{r}\right)d\vec{r}. \tag{2}$$

The kinetic energy of non-interacting electron gas with density $\rho\left(\vec{r}\right)$ is represented by $T_s\left[\rho\left(\vec{r}\right)\right]$ as follows:

$$T_s\left[\rho\left(\vec{r}\right)\right] = -\frac{1}{2}\sum_{i=1}^{N} \int \psi_i^*\left(\vec{r}\right)\nabla^2\psi_i\left(\vec{r}\right)d\vec{r}, \tag{3}$$

and Eq. (2) is defined as exchange-correlation energy functional $E_{XC}[\rho]$. Introducing a normalization constraint on the electron density, $\int \rho\left(\vec{r}\right)d\vec{r} = N$, we get:

$$\frac{\delta}{\delta\rho\left(\vec{r}\right)}\left[E\left[\rho\left(\vec{r}\right)\right] - \mu\int \rho\left(\vec{r}\right)d\vec{r}\right] = 0, \tag{4}$$

$$\Rightarrow \frac{\delta E\left[\rho\left(\vec{r}\right)\right]}{\delta\rho\left(\vec{r}\right)} = \mu. \tag{5}$$

The above equation can be rewritten in terms of an effective potential, $V_{eff}\left(\vec{r}\right)$:

$$\frac{\delta T_s\left[\rho\left(\vec{r}\right)\right]}{\delta\rho\left(\vec{r}\right)} + V_{eff}\left(\vec{r}\right) = \mu, \tag{6}$$

where

$$V_{eff}\left(\vec{r}\right) = V_{ext}\left(\vec{r}\right) + \int \frac{\rho\left(\vec{r}'\right)}{|\vec{r}-\vec{r}'|}d\vec{r}' + V_{XC}\left(\vec{r}\right) \text{ and } V_{XC}\left(\vec{r}\right) = \frac{\delta E_{XC}\left[\rho\left(\vec{r}\right)\right]}{\delta\rho\left(\vec{r}\right)} \tag{7}$$

The V_{XC} is the exchange-correlation potential, then the one electron SchrÖdinger equation takes the form written as follows:

$$\left(-\frac{1}{2}\nabla_i^2 + V_{eff}\left(\vec{r}\right) - \in_i\right)\psi_i\left(\vec{r}\right) = 0 \tag{8}$$

So, solving $\rho\left(\vec{r}\right)$ we get:

$$\rho\left(\vec{r}\right) = \sum_{i=1}^{N}\left|\psi_i\left(\vec{r}\right)\right|^2, \tag{9}$$

The self-consistent solution is required due to the dependence of $V_{eff}\left(\vec{r}\right)$ on $\rho\left(\vec{r}\right)$. Calculations of electronic structures are generally approximated through local density approximations or generalized gradient approximations [30].

2.2 Approximations to exchange-correlation functional

The exchange-correlation functionals used in DFT are categorized into two namely; the local density approximation (LDA) [31] and the generalized gradient approximation (GGA) [32]. These functionals are discussed in sections 2.2.1 and 2.2.2.

2.2.1 Local density approximation

The local density approximation (LDA) is an approximation in which the exchange-correlation (XC) energy function depends upon the value of the electronic density at each point in density functional theory (DFT) [31]. It was first discovered by Kohn and Sham in the context of DFT which can be expressed as:

$$E_{XC}[\rho(r)] \cong \int dr\rho(r)\varepsilon_{XC}(\rho(r)), \tag{10}$$

where $\varepsilon_{XC}(\rho)$ is the exchange-correlation energy per electron in a uniform electron gas of density n [31]. In the uniform electron gas, electrons are distributed in interacting systems with an arbitrary spatial density ρ which acts as a parameter. It has been demonstrated that LDA delivers accurate results even if the electron density in the system is not gradually varied [31]. The function $\varepsilon_{XC}(\rho)$ is a combination of exchange and correlation contributions of $\varepsilon_{XC}(\rho) = \varepsilon_X(\rho) + \varepsilon_C(\rho)$. It is possible to calculate the exchange energy per particle of a uniform electron gas as follows:

$$\varepsilon_{XC}(\rho) = c_x \rho^{1/3}, \tag{11}$$

where $c_x = -(3/4)(3/\pi)^{1/3}$.

2.2.2 Generalized gradient approximation

The GGA is known to be semi-local approximation which means that the function does not use the local density $\rho(r)$ value but its gradient $\nabla\rho(r)$. Perdew and Wang [32] developed GGA which improves the total energies, atomization energies, energy barriers, and also the difference in structural energies. GGA takes the form:

$$E_{XC}^{GGA}[\rho] = \int (\rho(r), \nabla\rho(r)) dr, \tag{12}$$

The spin-independent form is considered in GGA but practically functional is more generally formulated in terms of spin densities $(\rho\uparrow, \rho\downarrow)$ and their correspondence gradients of $(\nabla\rho\uparrow, \nabla\rho\downarrow)$.

There are several GGA-based functionals that are the PBE [33], PBEsol [34], RPBE [35], BLYP [36], and AM05 [37]. PBEsol functional is a simple modification of PBE that differs only with two parameters. It is designed to improve the equilibrium properties of bulk solids and their surfaces of PBE in physics and surface science communities. The revised version of the PBE, such as the RPBE functional, is widely used in catalysis to improve the performance of PBE. In the case of AM05, it gives the best performance for applications of catalysis. The GGA-BLYP functional is widely used in the chemistry environment. Other known GGA-based functionals are meta-GGA [38], hyper-GGA, and generalized random-phase approximation. An extension of the GGA, the meta-GGA uses the kinetic energy density and its gradient as inputs to the function and gradient along with the functional density. Hyper-GGA offers an accurate treatment of correlation that goes beyond the level of LDA or GGA when using exact exchange (EXX) to deal with exchange correlation. The generalized random-phase approximation uses EXX and exact partial correlation.

In this chapter, the GGA-PBE [33] functional was used to optimize the $Ti_{50}Pd_{50-x}M_x$ systems as it provides accurate parameters for this material.

2.3 Computational code and implementation

2.3.1 VASP code

The Vienna Ab initio Simulation Package (VASP) code [39] was used to calculate structural, thermodynamic, electronic, and mechanical properties of ternary $Ti_{50}Pd_{50-x}M_x$ (M = Ru, Co, Os) alloys. VASP [39] is a computer program for carrying out ab initio quantum mechanical calculations by making use of a plane wave basis along with pseudopotentials or projector-augmented-wave (PAW) [40]. VASP can compute an approximate solution by solving the Kohn-Sham equations within DFT. In VASP, central quantities, such as one-electron orbitals, electron change density, and local potential, are expressed in the form of plane-wave basis sets. The ultra-soft pseudopotentials (US-PP) [41] or the PAW method [40] is used to describe the interactions between the electrons and an ion in VASP. The US-PP method (and the PAW method) are effective in reducing the number of plane waves per atom in transition metals and first-row elements. The code consists of two main loops namely: the outer and inner loop, where the outer loop optimizes the charge density while the inner loop optimizes the wavefunction. VASP code uses a wide range of exchange-correlation functionals such as LDA and GGA as well as Meta- and hyper-GGA and hybrid functionals. All functionals found in VASP have spin-degenerate and also spin-polarized versions.

2.3.2 Implementation

A convergence test was done before calculating any properties, this is to ensure that proper convergence is attained. As such precision was set at "accurate" to minimize errors in the calculation. Importantly, the structures were subjected to full geometry optimization (by allowing both lattice parameters and volume to vary) until the atomic forces were less than 0.01 eV/Å for the unit cell. This was done in order to prepare the structures to be at their ground state energy before determining any properties, such as elastic constants and electronic structures. The effects of exchange-correlation interaction are treated with the generalized gradient approximation (GGA) [29] of Perdew-Burke = Ernzerhof (PBE) [33] and were used with the PAW potential [40]. The strain value of 0.005 was chosen for the deformation of the lattice when calculating elastic properties. A plane-wave cutoff energy of 500 eV and a k-spacing of 0.2 were found to converge the total energy of the systems.

The input structure has a B2 phase with the space group Pm-3 m (as shown in **Figure 1a**, it is cubic and stable at high temperature. The positions of B2 atoms are

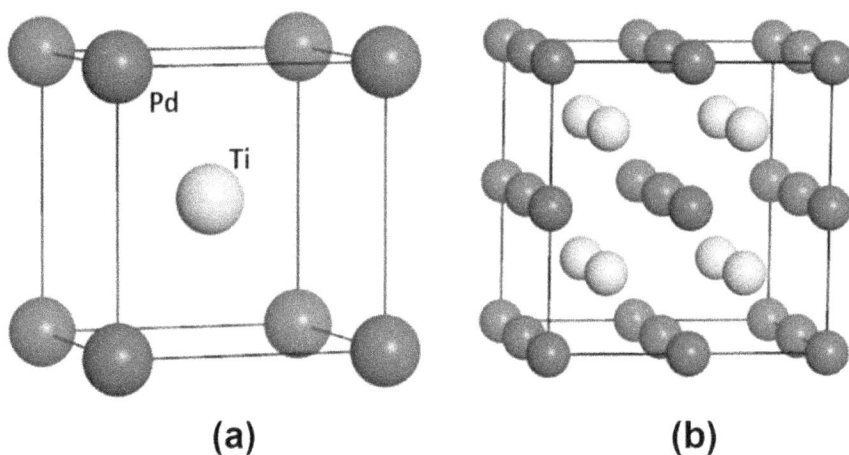

(a)   **(b)**

Figure 1.
(a) The B2 $Ti_{50}Pd_{50}$ structure with 2 atoms per unit cell and (b) a 2×2×2 supercell with 16 atoms per unit cell structures.

denoted by the Pearson symbol cP2 and the prototype is CsCl with all angles being 90°. B2 TiPd experimental observed unit cell parameters are a = b = c = 3.180° A [4]. In the case of ternary $Ti_{50}Pd_{50-x}M_x$, the calculations were carried out using a 2×2×2 supercell with 16 atoms (as shown in **Figure 1b**). The substitutional search tool embedded within the Medea software platform was used to substitute Pd with Ru, Os, and Co atoms, which provided the most stable compositions.

2.4 Theoretical background on calculated properties

2.4.1 Heat of formation

The heat of formation (ΔH_f) plays a crucial role in determining the thermodynamic stability of the different crystal structures where one can evaluate if the desired alloy can form or not beforehand. It is calculated according to Hess's law, which states that the standard enthalpy of an overall reaction is the sum of the standard enthalpies of individual reactions into which the reaction may be divided [42]. It provides insight into phase diagram construction and stabilities. The heat of formation is estimated by the following expression:

$$\Delta H_f = E_C - \sum_i x_i E_i, \qquad (13)$$

where E_C is the total energy of the respective structure calculated using first principle and E_i is the calculated total energy of the element i in the compound. A structure is considered stable if the value of the heat of formation is negative ($\Delta H_f < 0$) otherwise it is unstable. The heat of formation was used to determine the stability trend of $Ti_{50}Pd_{50-x}M_x$ alloys (M = Ru, Co, Os) as shown in Section 3.1.

2.4.2 Density of states

The density of states (DOS) can be used to predict the electronic stability of metal alloys. It is described by a function, g (E), as the number of electrons per unit volume and energy with electron energies near E. At a specific energy level, a high DOS means many states are open for occupation. In the case of states with DOS of zero, there is no state that can be occupied.

The electrical behavior of a material is determined by the location of E_f within the DOS. Metal alloys' stability can be predicted using the density of states (DOS). Any material's electronic density of states can be viewed as a qualitative measure of its electronic structure. DOS is then calculated as the sum of atomic contributions. The DOS is calculated by using the following expression:

$$n(\varepsilon) = 2\sum_{n,k}\delta(\varepsilon - \varepsilon_n^k) = \frac{2}{V_{BZ}}\sum_n \int \delta(\varepsilon - \varepsilon_n^k)dk \qquad (14)$$

where δ is the Dirac delta function and the k is integral extends over the BZ. The number of the electron in the unit cell is given by:

$$\int_{-\infty}^{\varepsilon f} n(\varepsilon)d\varepsilon. \qquad (15)$$

2.4.3 Mechanical properties

The elastic constants (C_{ij}) contain information regarding the strength of the materials against an externally applied strain and also act as stability criteria to study structural transformations from ground-state total-energy calculations. For a structure to exist in a stable phase, certain relationships must be observed between the elastic constants. There are various criteria established to deduce the mechanical stability of crystals for different lattice crystals. Accuracy in determining the elasticity of a compound is vital in understanding its mechanical stability and elastic properties. The B2 cubic crystal system has the simplest form of a stiffness matrix, with only 3-independent elastic constants c_{11}, $c_{12,}$ and c_{44} [43]. The mechanical stability criteria as outlined in Ref. [43] for the B2 Cubic and its alloyed structures can be expressed as follows:

$$c_{44} > 0; c_{11} > c_{12} \text{ and } c_{11} + 2c_{12} > 0 \qquad (16)$$

The stability criterion for the elastic constants must be completely satisfied for the structure to be stable. The positive $C' = ((1/2(c_{11}-c_{12}) > 0)$ indicates the mechanical stability of the crystal, otherwise, it is unstable.

3. Results and discussion

3.1 Structural and thermodynamic properties

In **Figure 2**, the calculated equilibrium lattice parameters for the B2 $Ti_{50}Pd_{50-x}M_x$ (M = Os, Ru, Co) systems are shown. It is observed that the partial substitution of Pd

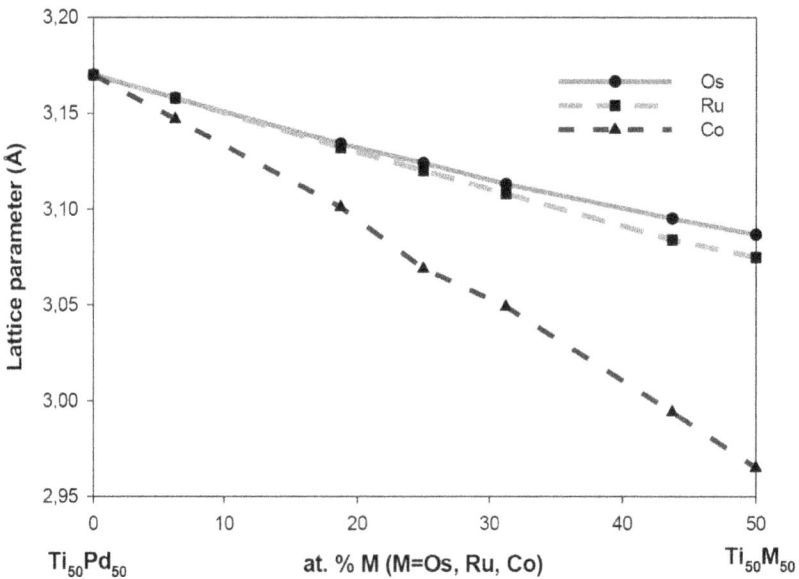

Figure 2.
The lattice parameter, a (Å) of the B2 $Ti_{50}Pd_{50-x}M_x$ (M=Os, Ru, Co) ($0 \leq x \leq 50$) ternary SMAs.

with Ru reduces the lattice parameters of the $Ti_{50}Pd_{50-x}Ru_x$ minimally (see **Figure 2**). This may be attributed to the small atomic radius of Ru as compared to that of Pd.

Furthermore, the lattice parameters of the $Ti_{50}Pd_{50-x}M_x$ system decrease as the Co and Os content is increased. This can be understood since the atomic radius of Pd is larger in size than Co, and Os. Recall that the lattice parameter of binary $Ti_{50}Pd_{50}$ was predicted to be 3.170 Å, which is larger than those calculated for the $Ti_{50}Pd_{50-x}M_x$ systems.

The heats of formation for the B2 $Ti_{50}Pd_{50-x}M_x$ systems are shown in **Figure 3**. As discussed in Section 2.4.1, the heat of formation is calculated to check the thermodynamic stability of the system. We observe that the ΔH_f decreases as Ru is increased this implies that the structure becomes stable at high Ru concentration (thermodynamically stable). Similar behavior was observed for Os as their values decrease with an increase in concentration, indicating thermodynamic stability. Furthermore, the addition of Co concentrations becomes less stable since the values of heat of formation increases as the content is increased. It is seen that Co substitution shows less stability, while the addition of Os and Ru enhances the stability of the $Ti_{50}Pd_{50}$ system at high concentration ($0 \leq x \leq 50$).

3.2 Electronic total density of states (tDOS)

In order to better examine the differences in electronic structures of $Ti_{50}Pd_{50-x}M_x$ alloys, it is important to analyze the behavior of the total density of states (DOS) near the Fermi level (E_f) with respect to the pseudogap. It is also known from previous studies that the DOS of structures of the same composition can be used to mimic the stability trend with respect to their behavior at the E_f [17, 44, 45]. A structure is considered the most stable if it has the lowest density of state at E_f. The DOS is expressed as the number of states per atom per energy interval.

Figure 4 shows the total DOS for $Ti_{50}Pd_{50-x}Ru_x$ alloys. The structures show an overlapping peak from the valence band (VB) to the conduction band (CB) suggesting a

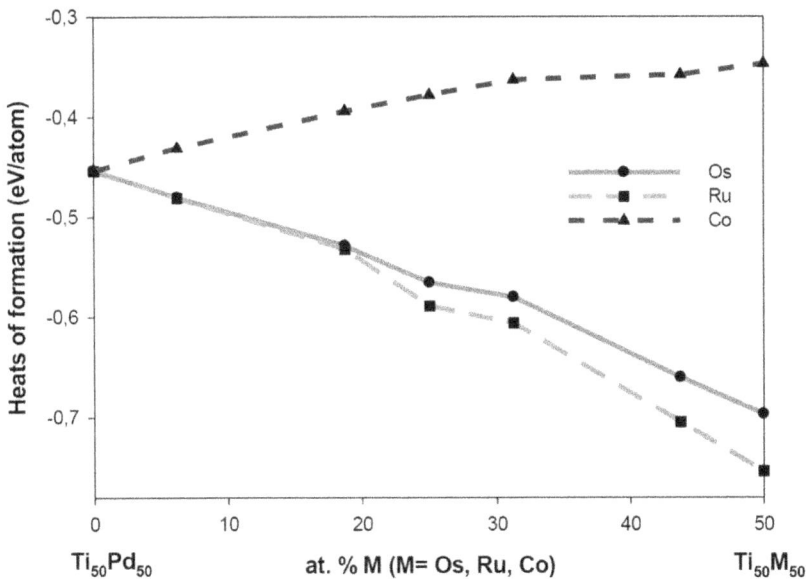

Figure 3.
Heats of formation of the B2 $Ti_{50}Pd_{50-x}M_x$ (M=Os, Ru, Co) ($0 \leq x \leq 50$) ternary SMAs.

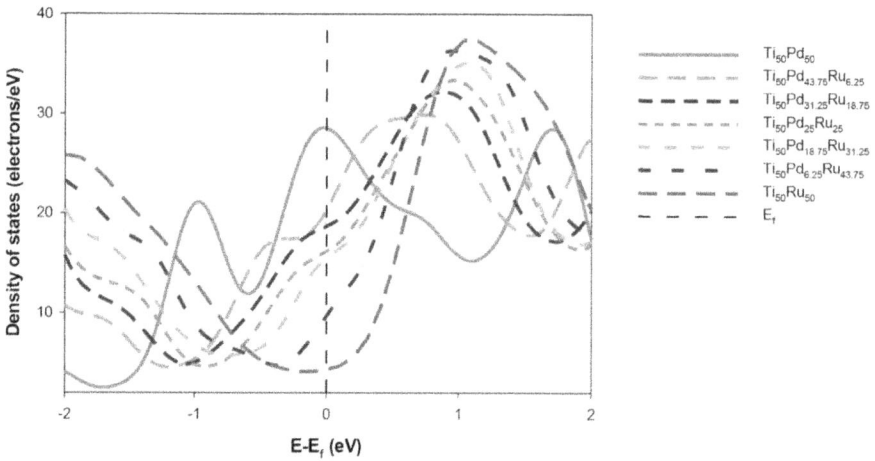

Figure 4.
Comparison of the total density of state for $Ti_{50}Pd_{50-x}Ru_x$ systems ($0 \leq x \leq 50$) against energy. The Fermi level is taken as the energy zero ($E-E_f = 0$).

metallic behavior since there is no visibility of a bandgap. A shift of DOS is also observed when Ru is added toward CB. As the composition of Ru is added, the pseudogap moves toward the E_f, indicating that the electronic stability is enhanced in particular for composition above 20 at. % Ru. It is noted that $Ti_{50}Ru_{50}$ has the lowest DOS at the E_f, which suggests that it is the most stable, while the $Ti_{50}Pd_{43.75}Ru_{6.25}$ is the least stable compared to other compositions. Furthermore, it was observed that the E_f coincides with the pseudogap. This observation suggests that $Ti_{50}Pd_{50-x}Ru_x$ is electronically stable at the high content of Ru. The predicted DOS analysis is consistent with the stability trend as predicted by the ΔH_f.

In **Figure 5**, we plot the total DOS for B2 $Ti_{50}Pd_{50-x}Os_x$ ($0 \leq x \leq 50$) alloys. As Os content is added, the DOS for $Ti_{50}Pd_{43.75}Os_{6.25}$ hits the peak at the shoulder near the E_f. Furthermore, at above 18.75 at. % Os, the pseudogap moves toward the E_f, which

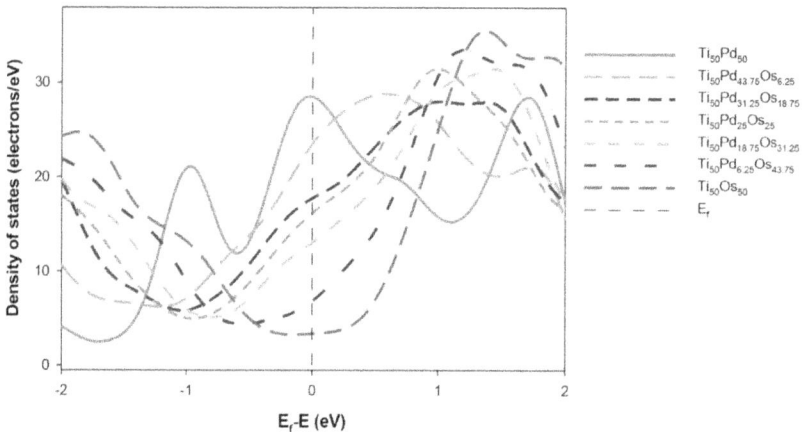

Figure 5.
Comparison of the total density of state for $Ti_{50}Pd_{50-x}Os_x$ systems ($0 \leq x \leq 50$) against energy. The Fermi level is taken as the energy zero ($E-E_f = 0$).

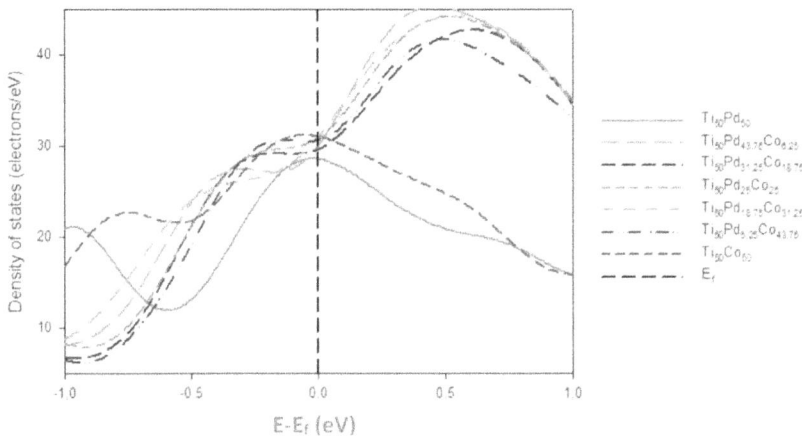

Figure 6.
Comparison of the total density of state against energy for Ti$_{50}$Pd$_{50-x}$Co$_x$ systems (0 ≤ x ≤ 50). The Fermi level is taken as the energy zero (E–E$_f$ = 0).

may suggest that the system starts to stabilize. It is also observed that at 50 at. % Os (Ti$_{50}$Os$_{50}$) has the lowest DOS near the E$_f$. The result suggests that Ti$_{50}$Pd$_{50-x}$M$_x$ is electronically stable at higher content of Os, consistent with the predicted ΔH$_f$.

Figure 6 shows and compares the calculated tDOS for Ti$_{50}$Pd$_{50-x}$Co$_x$ alloys. It is noted that as the Co content is increased the DOS hit the shoulder of the peak at E$_f$. The highest DOS along the E$_f$ is observed at the high composition of Co (50 at. %), while 6.25 at. % Co has the lowest DOS peak. This observation indicates that Co is not preferable to enhancing the electronic stability of Ti$_{50}$Pd$_{50}$ in good agreement with the predicted ΔH$_f$.

3.3 Elastic properties

The elastic constants (C$_{ij}$) are important parameters, which can be used to predict mechanical stability. To further investigate the stability of Ti$_{50}$Pd$_{50-x}$M$_x$, elastic constants were calculated. The stability criterion for the elastic constants must be satisfied for the structure to be defined as stable. The stability conditions for the cubic system are outlined in Section 2.4.3.

In **Figure 7**, the calculated elastic constants of the Ti$_{50}$Pd$_{50-x}$Ru$_x$ alloys (0 ≤ x ≤ 50) are shown. Recall that in order for the structure to be stable, it must satisfy certain stability criteria as discussed in Chapter 4. The positive C' ((1/2(c$_{11}$-c$_{12}$) > 0) indicates the mechanical stability of the crystal, otherwise, it is unstable. As indicated in Chapter 1, the binary B2 Ti$_{50}$Pd$_{50}$ alloy is mechanically unstable at 0 K due to negative elastic shear modulus (C' = − 5.37 GPa). So, the addition of a third element has been suggested in order to stabilize the B2 Ti$_{50}$Pd$_{50}$ alloy. In this case, the addition of Ru shows that the elastic constants c$_{11}$, c$_{12}$, and c$_{44}$ are positive for the entire concentration range (0 ≤ x ≤ 50) **Figure 7**. The c$_{11}$ and c$_{44}$ increase with the addition of Ru content while c$_{12}$ decreases suggesting that the structure is becoming mechanically stable (satisfying the stability condition, c$_{11}$ > c$_{12}$).

However, the predicted C$_{ij}$ does not satisfy the stability criteria for Ti$_{50}$Pd$_{50-x}$Ru$_x$ (when x = 6.25 and 18.75) since c$_{11}$ is less than c$_{12}$ which resulted in negative elastic shear modulus (C' < 0). Furthermore, it is observed that the elastic constants satisfy

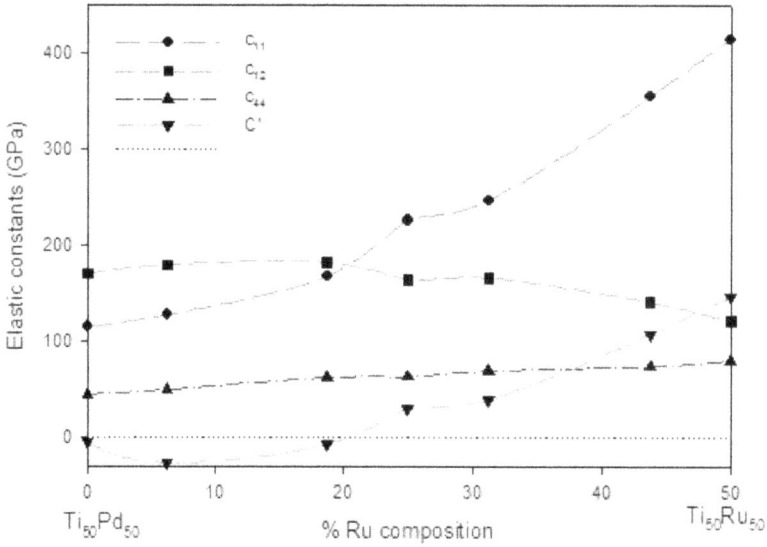

Figure 7.
The elastic constants (GPa) as a function of the atomic % Ru composition of $Ti_{50}Pd_{50-x}Ru_x$ SMAs.

the stability criterion above 20 at. % Ru indicating elastic stability (since $C' > 0$) of $Ti_{50}Pd_{50-x}M_x$ alloys. This may suggest that the addition of Ru reduces the martensitic transformation temperature of the TiPd due to an increase in C' above 20 at. % Ru. This observation has also been discussed in the previous study [46].

The calculated elastic properties of the $Ti_{50}Pd_{50-x}Os_x$ alloys ($0 \leq x \leq 50$) are also shown in **Figure 8**. It is noted that all the independent elastic constants c_{11}, c_{12}, and c_{44}

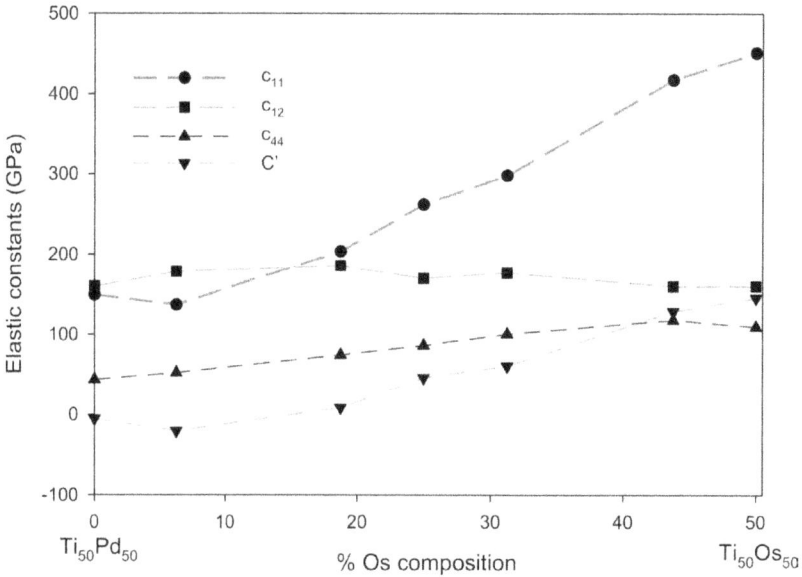

Figure 8.
The elastic constants (GPa) as a function of the atomic % Os composition of $Ti_{50}Pd_{50-x}Os_x$ SMAs.

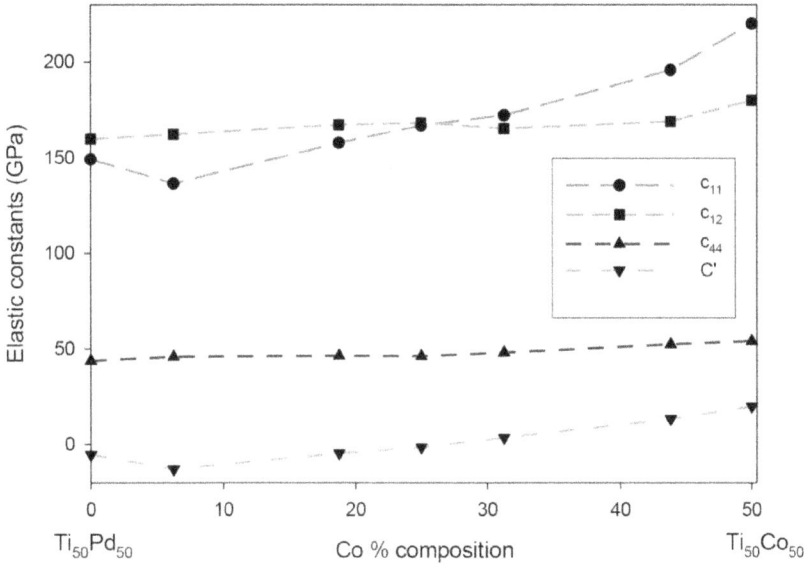

Figure 9.
Elastic constants (GPa) as a function of atomic percent Co for $Ti_{50}Pd_{50-x}Co_x$ where $0 \leq x \leq 50$.

are positive in the entire range of $Ti_{50}Pd_{50-x}Os_x$ alloys ($0 \leq x \leq 50$). It is observed that $c_{11} < c_{12}$ which resulted in negative C' below 18 at. % Os indicating mechanical instability at this concentration. Interestingly, the elastic constants satisfy the stability criterion above 18.75 at. % Os indicating elastic stability (since $C' > 0$) of $Ti_{50}Pd_{50-x}M_x$ alloys.

Figure 9 shows the comparison of elastic constant c_{11} and the c_{12} for the $Ti_{50}Pd_{50-x}Co_x$ at different concentrations. From the results, it is noted that the c_{11} is less than c_{12} below 25 at. % Co, which suggests instability at those compositions. It is clearly seen that the C' curve is below zero ($C' < 0$) at lower concentrations. However, the elastic constants satisfy the stability criterion above 31.25 at. % Co indicating elastic stability (since $C' > 0$) of $Ti_{50}Pd_{50-x}M_x$ alloys.

3.4 Anisotropy ratio

This section focuses on the anisotropy ratio to describe isotropic behavior and transformation as well as the ductility of the $Ti_{50}Pd_{50-x}M_x$ systems.

3.4.1 Isotropic and anisotropy behavior

It is important to study the elastic anisotropy of the systems in order to understand material properties and improve their mechanical durability. The anisotropy can be calculated as:

$$A = \frac{c_{44}}{C'} \qquad (17)$$

The factor $A = 1$ indicates that the material is isotropic and any value greater or smaller than 1 indicates an anisotropic behavior.

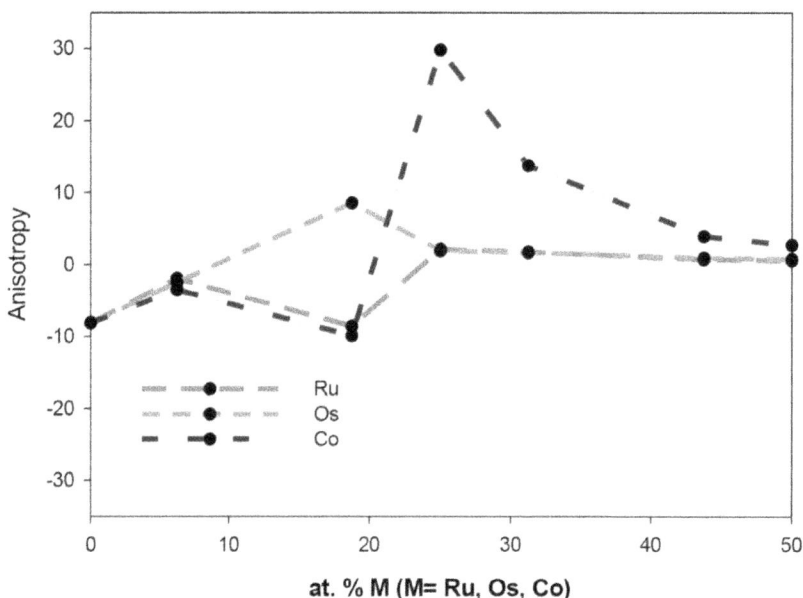

Figure 10.
Anisotropy as a function of atomic percent M (M=Os, Ru, Co) for $Ti_{50}Pd_{50-x}M_x$ where $0 \leq x \leq 50$.

The anisotropic plot depicts anisotropic behavior below 25 at. % for Ru, Os, and Co additions. However, A approaches unity ($A \approx 1$) for both Ru and Os between 25 and 50 at. % composition. These alloy systems have isotropic behavior at this composition range. The Co additions are highly anisotropic in the entire composition range (as shown in **Figure 10**).

3.4.2 Anisotropy and martensite transformation

An anisotropy ratio (A) plays an important role in predicting the martensitic transformation of the material. A martensitic transformation from B2 to B19 is observed when there is a higher value of A, while a good correlation between c_{44} and C' can be observed when A is smaller which leads to the transformation from B2 to B19′ [47].

In **Figure 10**, a higher value of A is observed at 25 at. % Ru, which suggests transformation from B2 to the B19 martensite phase. There is a coupling of c_{44} and C' at high concentration above 37 at. % Ru, which may suggest a possible phase transformation from B19 to the B19′ martensite phase. Thus, their addition results in a reversible martensitic transformation that is from B2 to B19, and B19 to B2. In the case of Os addition, A is higher for 18.75 at. % Os, which may suggest the transformation from B2 to B19 (see **Figure 10**). A strong coupling is observed between the c_{44} and C' at 43.75 at. % Os, which leads to the transformation from B19 to B19′ (since A is small). It can be concluded that $Ti_{50}Pd_{50-x}Os_x$ alloys transform from B2 to B19 (18.75 at. % Os) and then B19 transform into B19′ (43.75 at. % Os) due to a coupling of the c_{44} and C' at 0 K. This is a similar observation with TiNi and TiNi-based alloys [47]. In the case of Co addition, it was observed that A is negative below 18.75 at. % Co indicating that the martensite transformation is suppressed and the B2 phase is

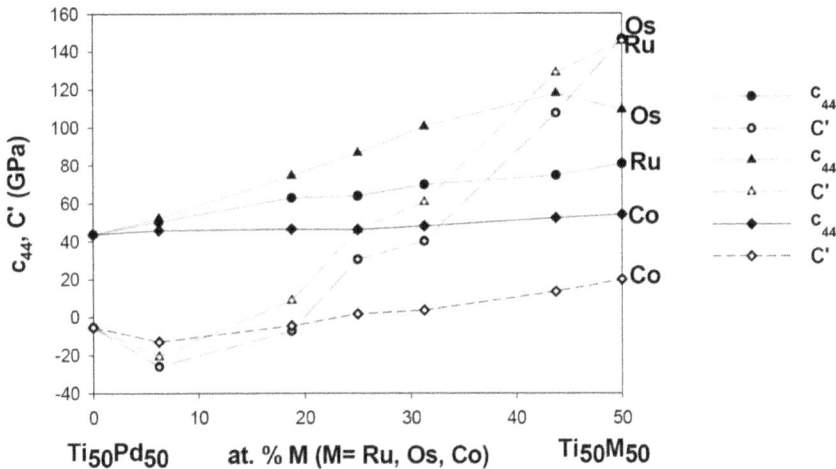

Figure 11.
The elastic constants c$_{44}$ and C' against the composition of Ti$_{50}$Pd$_{50-x}$M$_x$ alloys.

preserved (see **Figure 10**). As the composition of Co is increased to 31.25 at. %, A is higher than other compositions showing the transformation from B2 to B19. It can be deduced that B2 Ti$_{50}$Pd$_{50-x}$Co$_x$ alloys transform to B19 phase above 31 at. % Co.

Previously, Yi et al. [48] indicated that the c$_{44}$ and C' can be used to predict the change of martensitic transformation (Ms) at the Zener anisotropy factor A < 10. This factor measures the degree of anisotropy in solid and is calculated using Eq. (17). In a cubic crystal, C' is used to measure the basal-plane shear along the direction of {110} <1–10>, while the c$_{44}$ is along direction {001} <100> shear (non-basal plane shear) which is equal to {001} <1–10 > shear. Hence, the c$_{44}$ is playing an important role in controlling the transformation temperature of B2 to B19' [47]. Recall that the formation of the B19' phase is attributed to the coupling between c$_{44}$ and C' as proposed by Ren et al. [47]. Interestingly, as the Os is added the C' decreases and becomes negative below 6 at. % and positive above, which suggest stability (as shown in **Figure 11**). This observation indicates a possible increase in the transformation temperature below 6 at. % due to negative C'. It is noted that the entire Zener anisotropy factors are less than 10 for Os and Ru at. % composition, which indicates the possibility and reliability of the prediction of Ms. (see **Figure 10**). Furthermore, it was observed that the Zener anisotropy ratio is less than 10 for the addition of 6.25, 18.75, 43.75, and 50 at. % Co (except 25 and 31.25 at. % Co) as shown in **Figure 10**.

3.4.3 Anisotropy and ductility

The calculated A can also be used to check the ductility in metals. Thus, for a material to be considered ductile, the anisotropy ratio should be greater than 0.8 otherwise brittle [49]. An anisotropy values were found to be greater than 0.8 for 25 and 31.75 at. % Ru (**Figure 10**). The results imply that the alloy becomes ductile at this composition's ranges and brittle elsewhere. It was also found that the anisotropy ratio is greater than 0.8 above 18.75 at. % Os, which reveals ductile behavior. Furthermore, the Co addition is favorable above 25 at. % and 43.75 at. % (condition of ductility).

4. Conclusions

Ab initio DFT approach was used to study the equilibrium lattice parameters, heat of formation, and elastic properties of $Ti_{50}Pd_{50-x}M_x$ as potential HTSMAs. Our results of lattice parameters were found to be in good agreement to within 5% with the available experimental and theoretical values. Ternary alloying with Os, Ru, and Co were investigated. The results suggested that the addition of Ru and Os stabilizes the $Ti_{50}Pd_{50}$ structure since the heat of formation decrease with composition ($\Delta H_f < 0$). This was confirmed from the DOS analysis. It was found that the states are shifted at E_f as the concentration is increased. For example, as the composition of Ru is added, the pseudogap moved toward the E_f, indicating electronic stability especially above 20 at. % Ru. A similar trend was observed with Os addition.

The effect of ternary addition revealed that $Ti_{50}Pd_{50-x}Os_x$ alloys are mechanically stable above 18.75 at. % Os according to the criteria of mechanical stability. Furthermore, the C' was found negative below 25 at. % Ru ($C' < 0$, condition of instability) and becomes positive above this composition ($C' > 0$, condition of stability). The c_{11}, c_{44}, and C' for $Ti_{50}Pd_{50-x}Ru_x$ increase with an increase in Ru concentrations, while c_{12} decreases above 25 at. % Ru. It was found that the C' is negative below 31 at. % Co (instability characteristics) and becomes positive above this composition (condition of stability). This suggests that a possible HTSMAs material could be achieved above 31 at. % Co. This analysis has a direct impact on the transformation temperature. For example, a decrease in C' suggests that the Ms. is likely to increase.

An elastic anisotropy ratio (A) was used to describe the isotropic and anisotropic behavior of the $Ti_{50}Pd_{50-x}M_x$ systems. The analysis of A proves that the B2 $Ti_{50}Pd_{50-x}Co_x$ alloys displayed elastic anisotropy behavior (as A is less and greater than 1). It was found that A approaches unity ($A \approx 1$) for both Ru and Os between 25 and 50 at. % composition, suggesting isotropic behavior. Furthermore, we have also evaluated how alloying can impact the transformation temperature. The results suggest that the addition of Co, Ru, and Os on $Ti_{50}Pd_{50}$ alloy reduces the transformation temperature as indicated by positive C'. It was also found that the addition of Ru on $Ti_{50}Pd_{50}$ can result in transformation from B2 to B19 phase below 25 at. % Ru due to negative C'.

The ductile nature of $Ti_{50}Pd_{50-x}M_x$ alloys was confirmed by the value of anisotropy. It was revealed that increasing Os and Ru above 6.25 at. % could effectively improve the ductility of the compound. Furthermore, the anisotropy ratios were found greater than 0.8 above 18.75 at. % Os and Co, which reveal ductility behavior.

Acknowledgements

The computational work was carried out using computer resources at the Materials Modeling Center, University of Limpopo. The authors acknowledge the High-Performance Computing (CHPC) in Cape Town for their computing resources. The support of the South African Research Chair Initiative of the Department of Science and Technology is highly appreciated.

Author details

Ramogohlo Diale[1*], Phuti Ngoepe[2] and Hasani Chauke[2]

1 Advanced Materials Division, Mintek, Johannesburg, South Africa

2 Materials Modelling Centre, University of Limpopo, Sovenga, South Africa

*Address all correspondence to: ram@mintek.co.za

IntechOpen

References

[1] Jani JM, Leary M, Subi A, Gibson MA. A review of shape memory alloy research, applications and opportunities. Materials and Design. 2014;**56**:1078-1113

[2] Melton KN, Otsuka K, Wayman CM. General applications of SMA's and smart materials. In: Shape Memory Materials. Cambridge: Cambridge University Press; 1998. pp. 220-239

[3] Yamabe-Mitarai Y, Arockiakumar R, Wadood A, Suresh KS, Kitashima T, Hara T, et al. Ti(Pt, Pd, Au) based high temperature. Materials Today: Proceedings. 2015;**2**:S517-S552

[4] Donkersloot HC, Van Vucht JHN. Martensitic transformations in gold-titanium, palladium-titanium and platinum-titanium alloys near the equiatomic composition. Journal of Less Common Metals. 1970;**20**:83-91

[5] Solomon VC, Nishida M. Martensitic transformation in Ti-rich Ti–Pd shape memory alloys. Materials Transactions. 2002;**43**:908-915

[6] Hisada S, Matsuda M, Yamabe-Mitarai Y. Shape change and crystal orientation of B19 martensite in equiatomic TiPd alloy by isobaric test. Metabolism. 2020;**10**:3754

[7] Golberg D, Xu Y, Murakami Y, Otsuka K, Ueki T, Horikawa H. High-temperature shape memory effect in Ti50Pd50-xNix (x = 10, 15, 20) alloys. Intermetallics. 1995;**22**:241-248

[8] Nishida M, Hara T, Morizono Y, Ikeya A, Kijima H, Chiba A. Transmission electron microscopy of twins in martensite in Ti-Pd shape memory alloy. Acta Materialia. 1997;**45**:4847-4853

[9] Guo C, Li M, Li C, Du Z. A thermodynamic modeling of the Pd–Ti system. Calphad. 2011;**23**:512-517

[10] Dwight AE, Conner RA Jr, Downey JW. Equiatomic compounds of the transition and lanthanide elements with Rh, Ir, Ni and Pt. Acta Crystallographica. 1965;**18**:835-839

[11] Golberg D, Xu Y, Murakami Y, Morito S, Otsuka K. Characteristics of Ti50Pd30Ni20 high-temperature shape memory alloy. Intermetallics. 1995;**3**: 35-46

[12] Huang X, Karin M, Ackland J. First-Principles Study of the structural energetics of PdTi and PtTi. Physical Review B. 2003;**67**:024101-024107

[13] Yamabe-Mitarai Y. TiPd- and TiPt-based high-temperature shape memory alloys: A review on recent advances. Metabolism. 2020;**10**:1531-1552

[14] Otsuka K, Oda K, Ueno Y, Piao M, Ueki T, Horikawa H. The shape memory effect in a Ti50Pd50 alloy. Scripta Metallugica et Materialia. 1993;**29**: 1355-1358

[15] Diale RG, Modiba R, Ngoepe PE, Chauke HR. Density functional theory study of TiPd alloying with Os as potential high temperature shape memory alloys. IOP Conference Series in Material Science and Engineering. 2019; **655**:012042

[16] Bozzolo G, Mosca HO, Noebe RD. Phase structure and site preference behavior of ternary alloying additions to PdTi and PtTi shape-memory alloys. Intermetallics. 2007;**15**:901-911

[17] Mahlangu R, Phasha MJ, Chauke HR, Ngoepe PE. Structural, elastic and

electronic properties of equiatomic PtTi as potential high-temperature shape memory alloy. Intermetallics. 2013;**33**: 27-32

[18] Otsuka K, Ren X. Recent developments in the research of shape memory alloys. Intermetallics. 1999;**7**: 511-528

[19] Baldwin E, Thomas B, Lee JW, Rabiei A. Processing TiPdNi base thin-filmshape memory alloys using ion beam assisted deposition. Surface and Coating Technology. 2005;**200**:2571-2579

[20] Diale RG, Modiba R, Ngoepe PE, Chauke HR. The effect of Ru on Ti50Pd50 high temperature shape memory alloy: A first-principles study. MRS Advance. 2019;**4**: 2419-2429

[21] Diale RG, Modiba R, Ngoepe PE, Chauke HR. Phase stability of TiPd1-xRux and Ti1-xPdRux shape memory alloys. Materials Today: Proceedings. 2021;**38**:1071-1076

[22] Diale RG, Ngoepe PE, Chauke HR. Self-consistent charge density functional tight-binding (SCC-DFTB) parameterization and validation for Ti50Pd50-XRuX alloys. Computational Materials Science. 2023;**218**:111988

[23] Coutsouradis D, Davin A, Lamberigts M. Cobalt-based superalloys for applications in gas turbines. Materials Science and Engineering. 1987;**88**:11-19

[24] Haynes WM, editor. CRC Handbook of Chemistry and Physics. 92nd ed. Boca Raton: CRC Press; 2011: 2656

[25] Jahn'atek M, Levy O, Hart GLW, Nelson LJ, Chepulskii RV, Xue J, et al. Ordered phases in ruthenium binary alloys from high-throughput first-principles calculations. Physical Review B. 2011;**84**:214110-214,118

[26] Haynes WM, editor. CRC Handbook of Chemistry and Physics. 92nd ed. CRC Press;

[27] Kumar PK, Lagoudas DC, Zanca J, Lagouda MZ. Thermomechanical characterization of high temperature SMA actuators. Proceedings of SPIE. 2006;**6170**:306-312

[28] Hohenberg P, Kohn W. Inhomogeneous electron gas. Physical Review B. 1964;**136**:B864-B871

[29] Kohn W, Sham LJ. Self-consistent equations including exchange and correlation effects. Physical Review A. 1965;**140**:A1133-A1138

[30] Mattson AE, Schultz PA, Desjarlais MP, Mattsson TR, Leung K. Designing meaningful density functional theory calculations in materials science—A primer. Materials Science and Engineering. 2005;**13**:R1-R32

[31] Hedin L, Lundqvist BI. Explicit local exchange-correlation potentials. Journal of Physics C. 1971;**4**:2064-2082

[32] Perdew JP, Wang Y. Accurate and simple analytic representation of the electron-gas correlation energy. Physical Review B. 1992;**45**:13244-13,249

[33] Perdew JP, Burke K, Ernzerhof M. Generalized gradient approximation made simple. Physical Review Letters. 1996;**77**:3865-3868

[34] Perdew JP, Ruzsinszky A, Csonka GI, Vy-drov OA, Scuseria GE, Constantin LA, et al. Restoring the density-gradient expansion forexchange in solids and surfaces. Physical Review Letters. 2008;**100**:136406

[35] Hammer B, Hansen LB, Nørskov JK. Improved adsorption energetics within density-functionaltheory using revised Perdew-Burke-Ernzerhof functionals. Physical Review B. 1999;**59**:7413

[36] Peter M, Gill W, Johnson BG, Pople JA, Frisch MJ. The performance of the Becke-Lee-Yang-Parr (B-LYP) density functional theory with various ba-sis sets. Chemical Physics Letters. 1992;**197**:499

[37] Armiento R, Mattsson AE. Functional designed toinclude surface effects in self-consistent density functional theory. Physical Review B. 2005;**72**:085108

[38] Tao J, Perdew JP, Staroverov VN, Scuseria GE. Climbing the density functional ladder: nonempirical meta-generalized gradient approximation designed for molecules and solids. Physical Review Letters. 2003;**91**:146401

[39] Kresse G, Furthmüller J. Efficient Iterative Schemes for Ab-initio Total-energy Calculations Using a Plane-wave Basis Set. Physical Review B. 1996;**54**:11169-11,186

[40] Blöchl PE. Projector augmented-wave method. Physical Review B. 1994;**50**:17953-17,979

[41] Vandebilt D. Soft self-consistent pseudopotentials in a generalized eigenvalue formalism. Physical Review B. 1990;**41**:7892

[42] Masia M, Probst M, Rey R. Ethylene carbonate- Li+: A theoretical study of structural and vibrational properties in gas and liquid phases. The Journal of Physical Chemistry B. 2004;**108**:2016-2027

[43] Mehl MJ, Klein BM. First-principles calculation of elastic properties. In: Westbrook JH, Fleischer RL, editors. Intermetallic Compounds – Principles and Practice. Vol. 1. London: John Wiley and Sons, Ltd; 1994. pp. 195-210

[44] Gornostyrev YN, Kontsevoi OY, Maksyutov AF, Freeman AJ, Katsnelson MI, Trefilov AV, et al. Negative yield stress temperature anomaly and structural instability of Pt3Al. Physical Review B. 2004;**70**:014102

[45] Pankhurst DA, Nguyen-Manh D, Pettifor DG. Electronic origin of structural trends across early transition-metal disilicides: Anomalous behavior of CrSi2. Physical Review B. 2004;**69**:075113

[46] Tan CL, Cai W, Zhu JC. First-principles study on elastic properties and electronic structures of Ti-based binary and ternary shape memory alloys. Chinese Physics Letters. 2006;**23**:2863-2866

[47] Ren X, Miura N, Taniwaki L, Otsuka K, Suzuki T, Tanaka K, et al. Understanding the martensitic transformations in TiNi-based alloys by elastic constants measurement. Materials Science and Engineering A. 1999;**273–275**:190-194

[48] Zener C. Contributions to the theory of beta-phase alloys. Physics Review. 1947;**71**:846-851

[49] Gschneidner K et al. A family of ductile intermetallic compounds. Nature Materials. 2003;**2**:587-591

Chapter 4

Superelastic Behaviors of Molecular Crystals

Takuya Taniguchi

Abstract

Molecular crystals have medium mechanical properties between inorganic alloys and organic polymers. The material category of molecular crystals has recently shown unique mechanical responses induced by external stimuli such as light, heat, and force. This review explores the superelasticity of molecular crystals, a phenomenon first discovered by Takamizawa *et al*. in 2014. Molecular crystals can manifest superelasticity by much smaller stresses than typical shape memory alloys, reflecting weaker intermolecular interactions of molecular crystals. A novel photo-responsive occurrence of superelastic deformation was observed in a chiral salicylideneamine crystal, exhibiting photoisomerization and phase transition. This process, involving torsional bending and superelastic deformation within a single crystal, could offer new functionalities in photo-responsive materials. Furthermore, it was found that superelasticity is prevalent across the molecular space by an informatics approach. As data accumulate, materials informatics may unveil the underlying relationship between superelasticity and the structures of molecular crystals, potentially enabling innovative material design.

Keywords: superelasticity, molecular crystals, mechanical property, actuation, finite element analysis, materials informatics

1. Introduction

Superelasticity, the phenomenon of material's returning to the original shape even after large deformation beyond the elastic limit, has been observed only in some specific shape memory alloys. Superelasticity is a unique behavior unlike the normal elastic and plastic deformation occurring in all materials (**Figure 1a**). Shape memory materials deform irreversibly in the temperature range below the martensitic transition, and then the original shape is recovered upon heating due to the transition to the austenite phase. This phenomenon is known as the shape memory effect. On the other hand, in the temperature range above the martensitic transition temperature, even if a load beyond the elastic limit is applied to the material, the material of austenite phase deforms largely, and then immediately returns to its original shape upon unloading due to stress-induced martensitic phase (**Figure 1b**). This phenomenon is known as superelasticity. For example, nickel-titanium (NiTi) alloys are well-known as shape memory alloys,

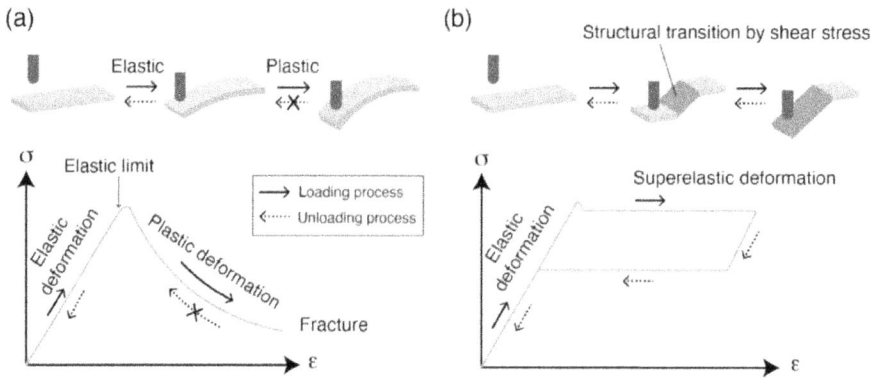

Figure 1.
Schematic illustration of the stress–strain curve of (a) typical elastic and plastic deformations and (b) superelastic deformation.

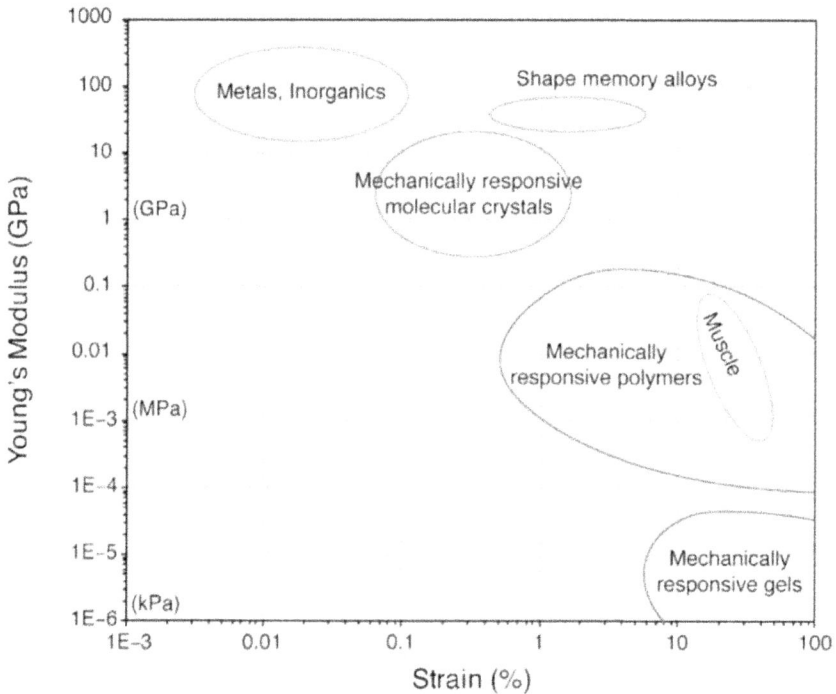

Figure 2.
Relationship between Young's modulus and strain of actuation materials. Reprinted with permission from [1]. Copyright 2019 John Wiley and Sons.

with elastic moduli ranging from 30 to 80 GPa and strain magnitudes from 0.5 to 5% (**Figure 2**) [2, 3]. Compared to soft organic polymers with a small modulus of 1 MPa ∼ 1 GPa and a large strain of 1 ∼ 100%, shape memory alloys are more rigid.

Molecular crystals have medium mechanical properties between those of hard inorganic materials and soft polymers (**Figure 2**) [1, 4]. The elastic modulus is about 1 ∼ 20 GPa, and the strain is about 0.1 ∼ 1% [5]. These mechanical properties originate from the ordered nature of crystals and the weak intermolecular interactions of organic molecules. The periodic arrangement of organic molecules in the structure leads to the ordered formation of intermolecular interactions, resulting in a large elastic modulus. On the other hand, the intermolecular interactions of organic molecules are composed of hydrogen bonds, π-π interactions, and van der Waals forces, so the interactions are smaller than those of inorganic materials [6]. Molecular crystals are attracting attention as novel stimuli-responsive materials called soft crystals because they take advantage of these moderate mechanical properties to undergo structural changes in response to relatively weak stimuli [7, 8]. For example, crystal deformations, such as bending, twisting, and jumping, appear upon light irradiation or temperature change [9–19]. The driving mechanism is the crystal structure change through photoisomerization or structural phase transition, and the deformation mode depends on the structural difference before and after the change. In addition, various types of crystals have been discovered, including elastic crystals that deform greatly under loading [20–26] and plastic crystals that quickly develop plastic deformation in a very narrow elastic region [27–30]. Furthermore, in this decade, superelastic phenomena, observed only in shape memory alloys, have been discovered in molecular crystals, attracting attention as a new type of shape memory material.

The superelasticity of molecular crystals is based on structural phase transition or twinning transition. These mechanisms also relate to other mechanical phenomena: ferroelasticity, actuation, shape-memory effect, and self-healing (**Figure 3**). The mechanical responses can be utilized for applications such as sensing, optics, soft robot, and catheter. Although the application to devices requires other specifications, such as the cost and the ease of integration with current technologies, researchers are trying to construct devices using molecular crystals, as explained in a later section.

This review briefly summarizes the superelastic phenomenon in molecular crystals. Various superelastic crystals have been found since their discovery in 2014 [31]. In addition, crystals that exhibit a unique phase transition, named the photo-triggered phase transition, exhibit superelasticity during crystal actuation. The last section discussed the potential applications of superelastic molecular crystals and the structural features of molecules that exhibit superelasticity in molecular crystals. As a prospect, materials informatics may construct strategies for designing superelastic molecular crystals.

2. Discovery of superelasticity in molecular crystals

The superelasticity of molecular crystals was discovered by Takamizawa *et al.* in 2014 [31]. The molecule terephthalamide (**Figure 4a**) yields plate-like crystals with

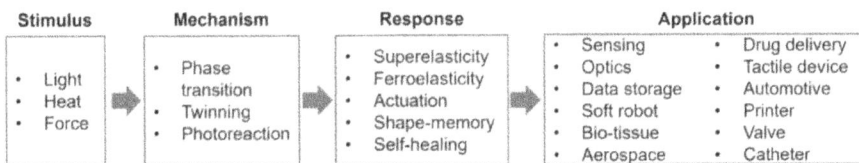

Stimulus	Mechanism	Response	Application	
• Light • Heat • Force	• Phase transition • Twinning • Photoreaction	• Superelasticity • Ferroelasticity • Actuation • Shape-memory • Self-healing	• Sensing • Optics • Data storage • Soft robot • Bio-tissue • Aerospace	• Drug delivery • Tactile device • Automotive • Printer • Valve • Catheter

Figure 3.
Typical mechanical responses of molecular crystals induced by external stimuli and examples of potential applications.

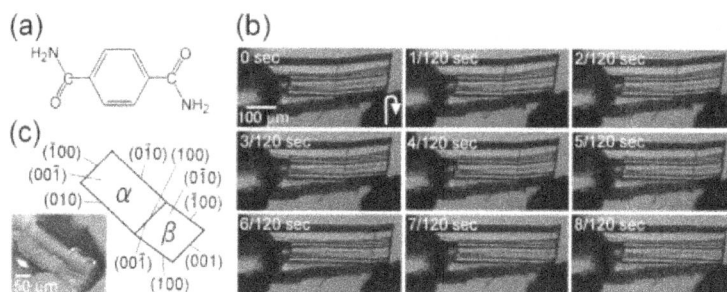

Figure 4.
Superelasticity of terephthalamide crystal. (a) the molecular structure of terephthalamide. (b) Pictures during superelastic deformation. The crystal is fixed at the left side, and loaded from the below right. (c) Assignment of crystal phases and face indices during superelastic deformation. Reprinted with permission from [31]. Copyright 2014 John Wiley and Sons.

the crystal structure of space group $P\bar{1}$. The crystal is in the α phase before loading, and when a load is applied to the (010) side face, the crystal deforms by creating a new crystal phase, as seen by the phase boundary (**Figure 4b**). Upon unloading, the new crystal phase gradually vanished, accompanying the movement of the phase boundary (**Figure 4b**). The new crystalline phase was assigned as the metastable β phase by X-ray diffraction measurement (**Figure 4c**). This molecular crystal was the first example of superelasticity.

Measuring the stress–strain curve during the crystal deformation is crucial to characterize the superelastic property. Takamizawa *et al.* measured the stress–strain curve and simultaneously observed crystal deformation under a polarized microscope (**Figure 5**). Initially, the platelet terephthalamide crystal fixed on a stand was the α phase. Upon loading, the β phase was created, as seen in different colors due to the change of molecular orientation and birefringence property (**Figure 5a**). During the observation, the time profile of the applied force was monitored, corresponding to snapshots of the crystal deformation (**Figure 5b**). The stress–strain curve was obtained from this measurement and displayed hysteresis of stress, indicative of superelasticity (**Figure 5c**).

This deformation is due to a structural phase transition: loading on the (010) plane of the α-phase induces shear stress, which induces a transition to the metastable β-phase. Comparing the superelasticity of terephtalamide molecular crystal and typical NiTi alloy is essential to characterize the differences in superelastic properties. The energy storage of terephtalamide crystal was 62 kJ m^{-3}, 226 times smaller than 14 MJ m^{-3} of typical NiTi alloy [31, 32]. This smaller energy storage of the molecular crystal is consistent with the smaller lattice energy compared with alloys. In other words, superelastic molecular materials can produce large deformations by small energy inputs. Thus, it was demonstrated that noncovalent interactions can produce superelasticity with great precision and can be controlled more precisely than in metallic alloys.

After this discovery, various superelastic molecular crystals have been found [33–40]. The primary mechanism of expressing superelasticity was based on mechanical twinning. Crystal twinning generally occurs during crystal growth. It occurs when two or more adjacent crystals of the same molecule are symmetrically oriented to share some of the same crystal lattice points, and two

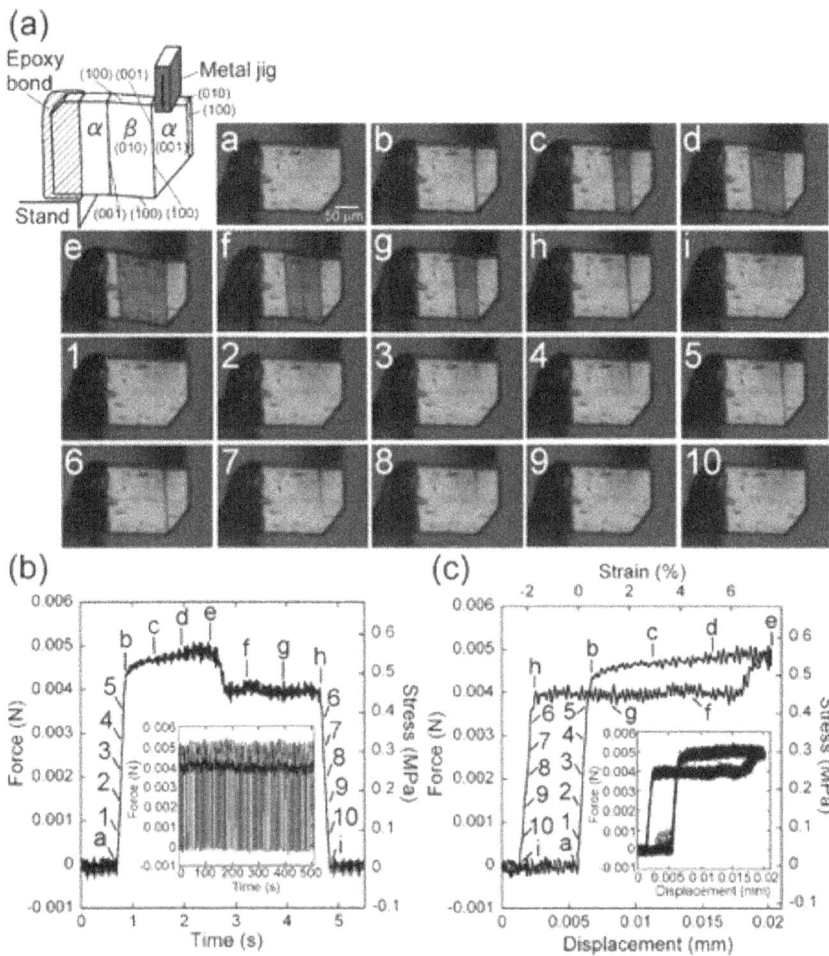

Figure 5.
(a) Snapshots of crystal deformation under a polarized microscope. Due to the simultaneous measurement, each snapshot was assigned to the time profile. (b) Time profile of applied force. Inset is the result of 100 cycles. (c) Stress–strain curve. Inset is the repetition result of 100 cycles. Reprinted with permission from [31]. Copyright 2014 John Wiley and Sons.

separate crystals grow tightly bound. The plane on which the twin lattice points are shared is called the twinning plane. Twinning is also manifested by shear stress on the crystal, which is the deformation mechanism of some superelastic molecular crystals (**Figure 6**). Sasaki summarized molecular structures which have been reported to manifest superelasticity or ferroelasticity induced by mechanical twinning [41]. Ferroelasticity is also the mechanism for large deformation like superelasticity, but the deformation does not return just by removing the load. Such ferroelasticity was often found in molecular crystals [42, 43]. **Figure 6** displays that there are diverse molecules that exhibit superplasticity and ferroelasticity. In addition, some molecular crystals manifest both effects by twinning mechanism.

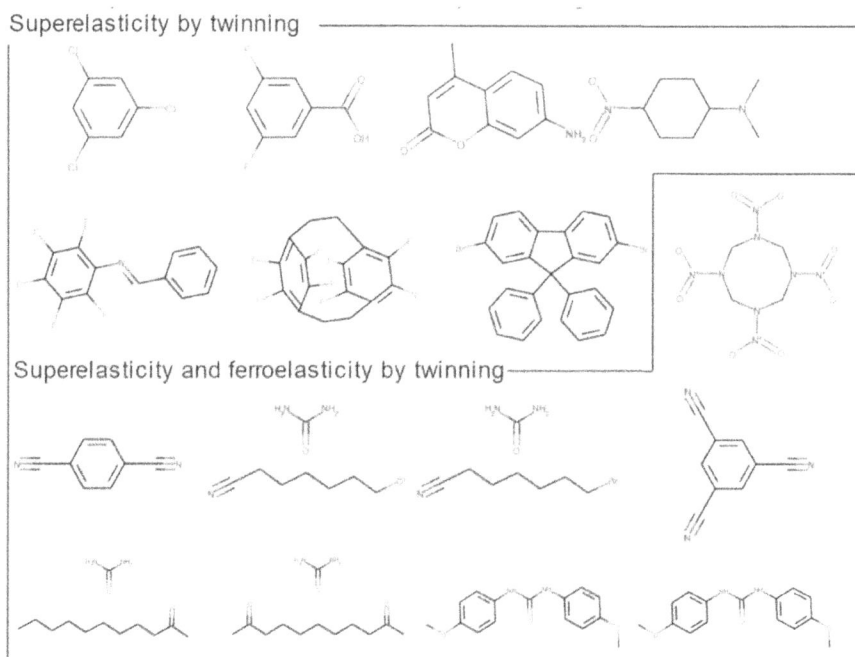

Superelasticity by twinning

Superelasticity and ferroelasticity by twinning

Figure 6.
Representative examples of superelastic molecular crystals by mechanical twinning.

3. Superelasticity of actuating crystal

3.1 Actuation by a photo-triggered phase transition

In the superelastic phenomena described in the previous section, a force is input, and superelastic deformation occurs as output. On the other hand, in actuation, energy other than force is the input, and force is the output extracted from material deformation. Thus, although superelasticity and actuation are essentially different phenomena, it was found that superelastic deformation occurs during actuation in chiral salicylideneamine crystal (**Figure 7a**) [44, 45].

Two relatively strong intermolecular interactions are formed in the plate-like crystal of the chiral salicylideneamine (**Figure 7b**): the CH-π interaction and the CH-O hydrogen bond. CH-π interactions are mainly formed by the face-to-face stacking of naphthyl rings along the *a*- and *b*-axes, and CH-O interactions are formed between the OH group of one molecule and the *tert*-butyl group of the adjacent molecule along the *b*-axis (**Figure 7c**). The rigid interaction layers are stacked along the *c*-axis via weaker van der Waals interactions between bulky *tert*-butyl substituents. The energy framework was calculated to evaluate the strength of this intermolecular interaction, and it was confirmed that the strong and weak interaction layers (**Figure 7c**). Also, when viewed from the (010) plane, the strong and weak interaction layers are arranged alternately; these layers are spread two-dimensionally along the *a*- and *b*-axes (**Figure 7d**). When viewed from the (001) plane, only the stronger interactions are seen due to the layer-by-layer stacking (**Figure 7e**).

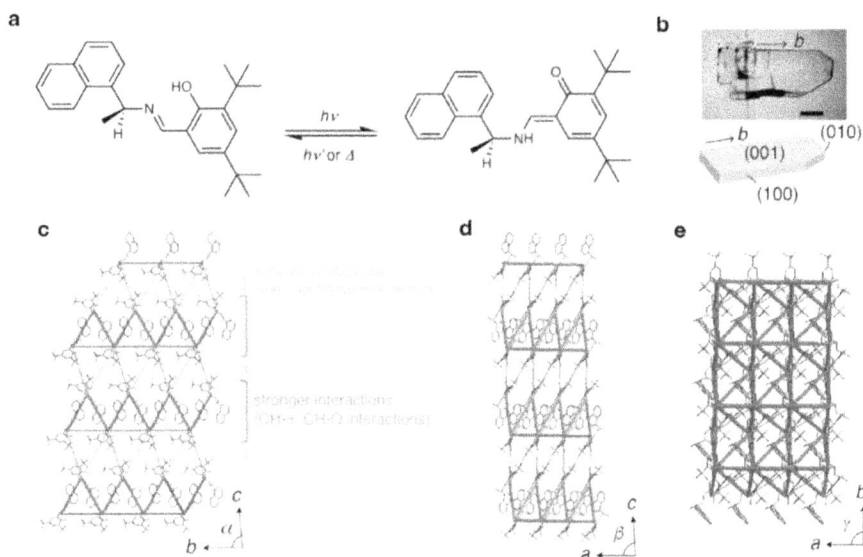

Figure 7.
(a) Molecular structure of chiral salicylideneamine. (b) Photograph and illustration of crystal shape and face index. Scale bar is 1 mm. (c-e) Molecular packing and energy framework, viewed from (c) (100) side face, (d) (010) cross-section face, and (e) (001) top face. Adapted from [44] under a creative commons attribution 4.0 international license (http://creativecommons.org/licenses/by/4.0/).

Light irradiation on the (001) plane of the crystal caused a unique phase transition called the photo-triggered phase transition (**Figure 8**). Ultraviolet (UV) light irradiation of the crystal induces photoisomerization from the enol to the *trans*-keto form, which generates stress in the crystal structure. When the generated stress accumulates to a certain amount, a phase transition of the crystal structure is induced to reduce the stress, and finally, a new crystal phase is formed (**Figure 8**). Here, it should be noted that photoisomerization occurs only in some parts of the bulk crystal near the irradiated surface, and a small number of the *trans*-keto form (approximately 5%) induce the phase transition of the entire crystal.

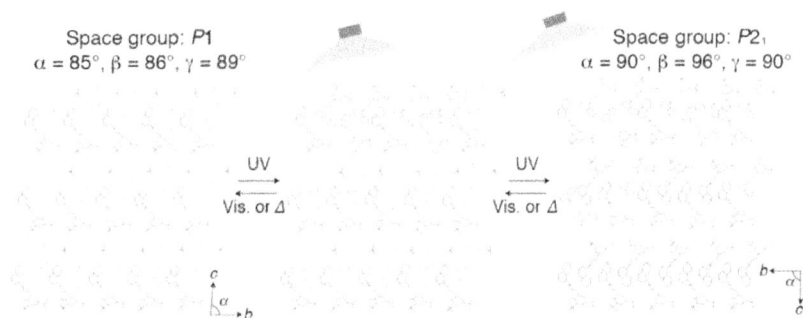

Figure 8.
Proposed mechanism of the photo-triggered phase transition and the measured lattice angles. Chiral salicylideneamine molecules in green and yellow reflect Z' = 2, and the trans-keto form is shown in red. Adapted from [44] under a creative commons attribution 4.0 international license (http://creativecommons.org/licenses/by/4.0/).

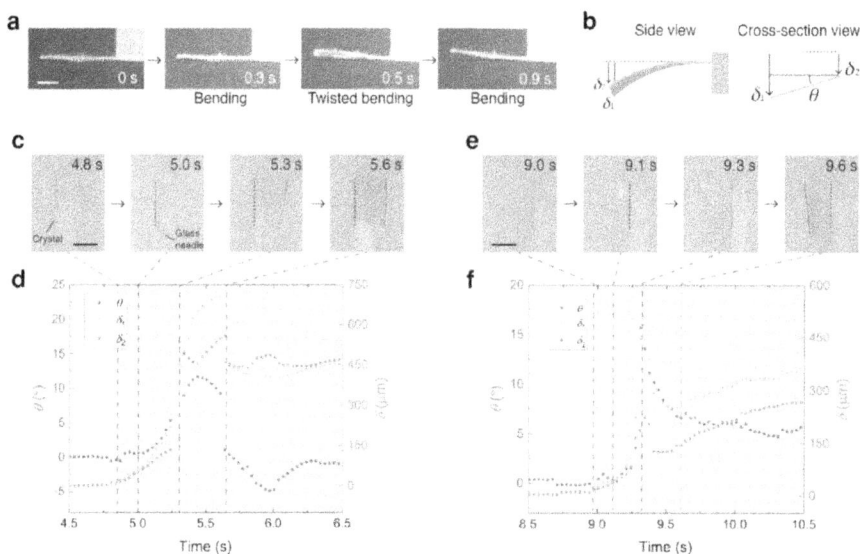

Figure 9.
(a) Photographs of typical deformation of the chiral salicylideneamine crystal, fixed on a glass plate, irradiated by UV light (365 nm). Scale bar is 1 mm. (b) Definition of torsion angle θ and displacements δ_1 and δ_2. The dotted lines are the initial position. (c) Cross-section view when irradiated on the (001) face, and (d) time profile of θ, δ_1, and δ_2. (e) Cross-section view when irradiated on (00$\bar{1}$) face, and (f) time profile of θ, δ_1, and δ_2. Scale bars in (c) and (e) are 0.5 mm. The regions highlighted in purple represent UV light at 180 mWcm^{-2}. Adapted from [44] under a creative commons attribution 4.0 international license (http://creativecommons.org/licenses/by/4.0/).

Because photoirradiation of the crystal causes photoisomerization and photo-triggered phase transitions, UV irradiation of the (001) plane leads to complicated crystal deformation. The deformation is divided into three steps (**Figure 9a**). The first step is a simple bending toward the light source due to photoisomerization. The second step is twisted bending due to the progression of the photo-triggered phase transition. The third stage is a simple bend toward the light source due to photoisomerization after the completion of the transition. Finally, when the light is stopped, the shape returns to its original shape reversibly.

A cross-section view can quantify complicated actuation (**Figure 9b**). Displacements at both edges are represented as δ_1 and δ_2, and the torsional angle as θ. Snapshots showed the actuation behavior observed from the cross-section view, and the time profile of displacements and torsional angle were measured (**Figure 9c-f**). Notably, the twisted bending finished within 1 second upon light irradiation, indicating the photo-triggered phase transition started and finished during that time.

3.2 Simulation of crystal deformation

Torsional bending is attributed to the photo-triggered phase transition, which arises during the initial photo process. For controlling the actuation behavior and understanding the actuation mechanism, it is crucial to simulate and replicate the material mechanics. Finite element analysis (FEA), commonly used for mechanical simulation, is suitable for the purpose and requires a simplified actuation model. The chiral salicylideneamine crystal displayed a stepwise actuation pattern: simple bending, torsional bending, and then simple bending again. Hence, it is considered a dynamic multilayer model comprising h_1 as the photoisomerization layer, h_2 as the transition layer, and h_3 as the remaining layer (**Figure 10a**). The crystal thickness is denoted as h_0. Considering light irradiation from the top surface, h_1 and h_2 are assumed to originate from the top surface. As the maximum photoproduct generation is 5%, h_1 can be much smaller than h_0, and the maximum value of h_2 can be equal to

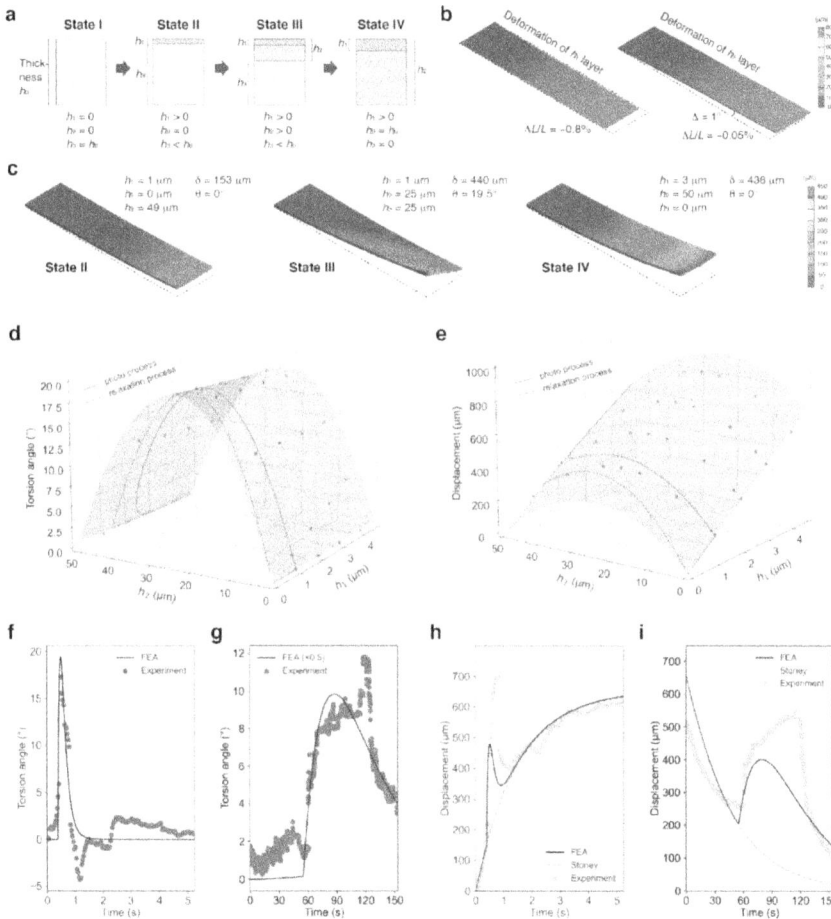

Figure 10.
FEA-simulation of crystal deformation. (a) Dynamic multi-layer model. (b) Independent assumed deformations of h_1 *and* h_2 *layers. The original dimensions of the plate object are 4.0mm in length and 0.94mm in width. Deformation is enhanced 6.4 times because raw deformation is much smaller than the object size. (c) Simulated typical deformations. (d, e) simulated dependence of (d) torsion angle and (e) maximum displacement on the thicknesses of* h_1 *and* h_2 *layers. Blue dots are the simulated points, and the response surfaces are drawn by a polynomial function. Red lines are the estimated route that reproduces the observed torsion angle and displacement. (f, g) comparison of the simulation and the torsion angle in the (f) photo-process and (g) relaxation process. (h, i) comparison of the simulation and the displacement in the (h) photo-process and (i) relaxation process. Reprinted from [44] under a creative commons attribution 4.0 international license (http://creativecommons.org/licenses/by/4.0/).*

h_0. FEA was conducted based on this model with the primary objective of reproducing simple and torsional bending. Once achieved, the effect of h_1 and h_2 thickness on deformation was examined to understand the progression of photoisomerization and the photo-triggered phase transition within the crystal.

For FEA, the material's dimensions were 4.0 mm in length, 0.96 mm in width, and 50 µm in thickness (h_0), approximately equivalent to the crystal shown in **Figure 9**. Then, pure effects of photoisomerization and photo-triggered phase transition were replicated: h_1 undergoes longitudinal shrinkage (-0.8%), and h_2 undergoes shear deformation ($\Delta 1°$) with slight shrinkage (-0.05%) (**Figure 10b**). These pure effects were replicated based on the results of X-ray diffraction measurements and implemented by thermal and piezoelectric effects because photo effects cannot be directly incorporated into FEA. Then, FEA successfully reproduced the simple bending, torsional bending, and consecutive simple bending by changing h_1, h_2, and h_3 (**Figure 10c**).

Then, the influence of h_1 and h_2 on deformation was evaluated. The values of h_1 and h_2 were systematically varied within the range of 0 to 5 µm and 0 to 50 µm, respectively. Polynomial regression was then employed to construct response surfaces illustrating the relationship between the torsion angle, maximum displacement, and the two thickness parameters (**Figure 10d, e**). Although this response function does not incorporate a time factor, assuming that h_1 is proportionate to the progression of the photoisomerization reaction, it can be expressed using the following equation

$$h_1(t) = h_{1,max} \left\{ 1 - exp\left(-\frac{t}{\tau_{1p}}\right) \right\} when\, t \geq 0 \tag{1}$$

Here, $h_{1,max}$ represents the maximum depth of the photoisomerization layer, t denotes the duration of light irradiation, and τ_{1p} represents the time constant. The progression of h_2 is also influenced by photoisomerization, which induces a phase transition. Therefore, it is posited that h_2 can be described by a similar exponential function, characterized by an independent time constant and a certain delay, t_{delay}, from $t = 0$.

$$h_2(t) = h_{2,max} \left\{ 1 - exp\left(-\frac{t}{\tau_{2p}}\right) \right\} (t \geq t_{delay}) \tag{2}$$

where $h_{2,max}$ denotes the maximum depth of the transition, and τ_{2p} represents the time constant. As $h_{2,max}$ can be equivalent to h_0, the optimization process involves determining the optimal values for the remaining parameters: $h_{1,max}$, τ_{1p}, τ_{2p}, and t_{delay}. To account for the relaxation process in the analysis, time-related factors are introduced under the assumption that h_1 and h_2 depend on the reverse chemical reaction as follows

$$h_1(t) = h_{1,max}\, exp\left(-\frac{t}{\tau_{1r}}\right)(t \geq 0) \tag{3}$$

$$h_2(t) = h_{2,max}\, exp\left(-\frac{t}{\tau_{2r}}\right)(t \geq t_{delay}) \tag{4}$$

where τ_{1r} and τ_{2r} represent the relaxation time constants. In Eq. (3) and (4), τ_{1r}, τ_{2r}, and t_{delay} are the parameters to be optimized, as the others have been determined in the photo process.

Following parameter optimization, the FEA-based simulations demonstrated a good fit with the observed torsion angle and displacement of the crystal during both the photo and relaxation processes (**Figure 10f–i**), albeit with some discrepancies in displacement. Notably, this simulation outperformed the classical Stoney's bimorph model in predicting displacement behavior (**Figure 10h,i**).

The optimized results shed light on the progression of the h_1 and h_2 layers within the crystal (depicted by the red lines in **Figure 10d, e**). During the photo-process, $h_{1,max}$ attained a value of 4.7 µm, while the h_2 layer commenced with a delay time of 0.38 s at a rate ten times faster than the h_1 layer, resulting in the maximum torsion angle occurring when $h_2 = h_0/2$. Subsequently, the h_2 layer reached $h_0 = 50$ µm, followed by a subsequent increase in the h_1 layer up to $h_{1,max}$. In the relaxation process, only h_1 initially decreased, and then h_2 began to decrease when h_1 reached approximately 1.5 µm. The relaxation speeds of h_1 and h_2 were found to be equal based on the fitting. Notably, there exists hysteresis between the photo and relaxation processes. Considering that photoproducts are responsible for inducing stress, which in turn triggers the phase transition, and that the release of stress without UV light leads to the reverse transition, this hysteresis can be considered analogous to the stress–strain curve of superelasticity. Thus, it can be concluded that superelasticity manifests in the photo-triggered phase transition due to the stress induced by photoproducts.

4. Other mechanical effects: shape memory and self-healing

Structural phase transition relates to superelasticity and shape memory effects of molecular crystals. It is reported that terephthalic acid crystals can undergo a mechanically induced phase transition without delamination upon bending while retaining their overall crystal integrity [46, 47]. Plastically bent crystals exhibit bimorph behavior, and their phase uniformity can be restored thermally by raising the crystal's temperature above the phase transition point. This thermal treatment recovers the original straight shape, and a reverse thermal treatment can induce shape memory effects similar to those observed in certain metal alloys and polymers. It is anticipated that similar memory and restorative effects are common in other molecular crystals with metastable polymorphs. This example insists on the advantage of utilizing intermolecular interactions to achieve mechanically adaptive properties in organic solids.

Self-healing of molecular crystals is less related to superelasticity but is crucial in the recovery of mechanical integrity even after appearing the crystal fracture. The first finding of self-healing in molecular crystals was discovered by Commins *et al.* using crystals of dipyrazolethiuram disulfide [48]. They broke a piece of single crystal into two pieces and then compressed it so that the cut surfaces of the two pieces touched each other. A healing degree of 6.7% was observed after mild compression of these crystals. It was hypothesized that the self-healing property of the material is attributed to the disulfide shuffling mechanism. The crystal structure has three close S − S contacts, which could form new bonds at the interfaces of two crystals.

Another mechanism also achieved self-healing. Bhunia *et al.* found that piezoelectric molecular crystals were autonomously self-healed in milliseconds with crystallographic precision [49]. This self-healing was very fast due to the unnecessity of long-time compression. The self-healing mechanism arises from stress-induced electrical charges on fracture surfaces, leading to the precise recombination of the broken pieces through an electrostatically driven, diffusionless self-healing process. Although the self-healing of molecular crystals has yet to encounter many examples, such crystals will improve the durability and robustness of devices if other self-healing crystals are found.

5. Prospects

5.1 Applications of superelastic molecular crystals

As briefly explained in the introduction, molecular crystals have various mechanical properties and may lead to applications. For example, superelastic and ferroelastic organic semiconductor crystals can be used in a flexible device [50, 51]. Flexible single-crystal electronic devices were achieved by leveraging the bending-induced ferroelastic transition of an organic semiconductor crystal. These devices can withstand strains of over 13% while preserving the charge carrier mobility of unstrained crystals. This advancement lays the groundwork for high-performance ultra-flexible single-crystal organic electronics, opening doors to their utilization in sensors, memories, and robotic applications. In addition, superelasticity in the actuating crystal may work for lifting and moving objects, photoswitches, and micromechanical devices.

Superelasticity and shape memory effect in nanoparticles also have the potential for future applications. Experimental studies have demonstrated that the shape memory properties of ceramics can be enhanced by reducing the density of grain boundaries. For example, Zhang *et al.* observed a superelastic strain of 8.3% in single-crystal nanoparticles of zirconia-based ceramics, which could be fully recovered by simply removing the compressive load [52]. However, they also confirmed that the shape memory behavior deteriorated as the number of loading-unloading-heating–cooling cycles increased. This degradation of shape memory properties was attributed to the formation of an amorphous phase, its accumulation around the load contact area, and an increase in surface roughness. If the decline in function of nanoparticles is controlled, nanoparticle-based materials may be implemented or replaced with the current technologies. A similar strategy using nanoparticles will be applied to molecular crystals to improve the superelasticity and other mechanical properties.

5.2 Materials informatics for molecular design

As mentioned above, many molecular crystals have been reported to exhibit superelasticity, but the theoretical science behind this phenomenon has not yet been established. Therefore, it is crucial to find the structural characteristics of the molecules in which superelasticity is observed.

In order to understand the structural characteristics of molecules, it is necessary to represent molecules mathematically and compare the closeness between molecules. Therefore, we collected information from the literature on molecules that exhibit superelasticity, ferroelasticity, thermal phase transitions, and mechanochromic luminescence in crystals. The mechanochromic molecular crystals are summarized in [53], and thermal phase transitions are summarized in [54]. Then, the molecular structures were vectorized by Mordred descriptor [55]. Each dimension of the Mordred vectors was standardized to have the mean 0 and variance 1 and then reduced to two dimensions by a data embedding method, uniform manifold approximation and projection (UMAP). The molecular dataset and Python code are available via [56].

The scatter plot shows that molecules exhibiting superelasticity, ferroelasticity, and thermal phase transition have similar distributions and are widely distributed in

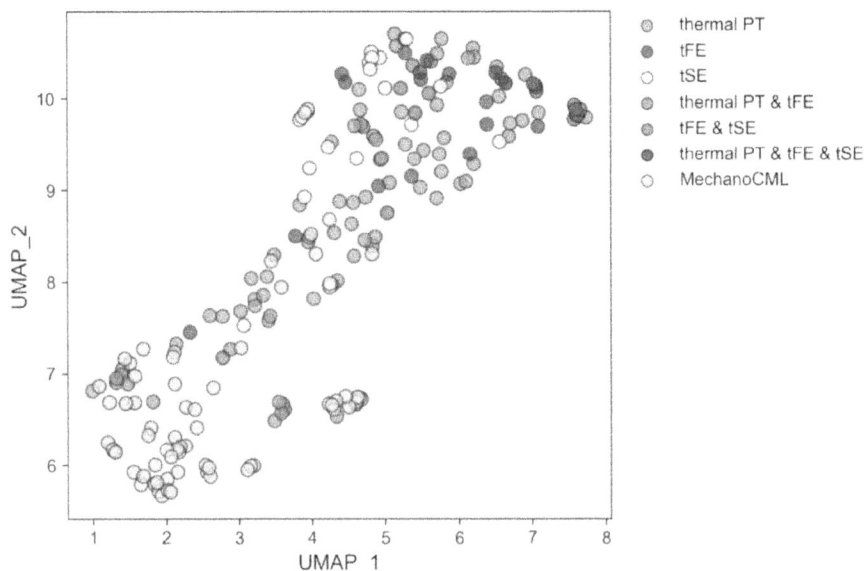

Figure 11.
Scatter plot of Mordred vectors embedded by UMAP. In the legend, thermal PT is a thermal phase transition, tFE is ferroelasticity by twinning, tSE is superelasticity by twinning, and mechanoCML is mechanochromic luminescence.

the 2-dimensional manifold space (**Figure 11**). This result indicates that molecules exhibiting superelasticity, ferroelasticity, and thermal phase transition are widely distributed in the material space with structural diversity. On the other hand, the molecules exhibiting mechanochromic luminescence are relatively tightly distributed, indicating that their molecular structures are similar. This difference can be attributed to the need for molecular structures involved in luminescence and that studies have been conducted with similar molecular structures with different substituents.

On the other hand, the relationship between superelasticity and molecular structure is still unclear. Therefore, it is necessary to find hidden patterns in the data, and this approach is called materials informatics (MI). Academia and many chemical and materials companies are developing materials utilizing MI, mainly focusing on industrially important inorganic solid materials and polymeric materials [57–59]. Although there are not many examples of MI research for molecular crystals compared to those materials, the number of reported cases is increasing, such as using machine learning for predicting polymorphism in pentacene crystals by Musil et al. [60]. As for superelasticity, if more data becomes available for machine learning, MI may be able to discover hidden relationships between molecular and crystal structures and develop them efficiently. Furthermore, MI can be applied to other mechanical properties, such as Young's modulus and maximum strain. Since materials' space is vast for human labor, MI-assisted materials design will benefit material research.

6. Conclusions

This review deals with the superelasticity of molecular crystals. The superelasticity of molecular crystals was discovered by Takamizawa *et al.* in 2014, and many superelastic and ferroelastic crystals have been discovered. It has been quantitatively evaluated that superelasticity develops at stresses as small as about 1/200th of NiTi, a typical shape memory alloy, showing the unique mechanical properties of molecular crystals. It was also found that superelastic deformation occurs during actuation in crystals in which photo-triggered phase transitions occur upon light irradiation. By observing and simulating the torsional deformation of chiral salicylideneamine crystal and clarifying the deformation mechanism, we have found that light-induced superelastic deformation occurs for the first time. This result is a unique photo-responsive phenomenon in which torsional deformation and superelastic deformation occur in a single crystal and may lead to new functionalities in photo-responsive materials. Other mechanical responses, shape memory, and self-healing, of molecular crystals were also described briefly. Finally, future applications and the possibility of materials informatics for molecular crystals were summarized. It was found that the molecules exhibiting superelasticity are widely distributed in the material space. As the number of data increases, the hidden relationship between superelasticity and molecular crystals and crystal structures can be captured by materials informatics and may be used for new material design.

Acknowledgements

The author thanks Mr. Daisuke Takagi and Mr. Ryo Fukasawa at Waseda University for curating molecular data for structural feature analysis. This study was financially supported by JSPS Grant-in-Aid (20H04677, 22 K14747).

Conflict of interest

The authors declare no conflict of interest.

Author details

Takuya Taniguchi
Waseda University, Tokyo, Japan

*Address all correspondence to: takuya.taniguchi@aoni.waseda.jp

IntechOpen

References

[1] Koshima H, Taniguchi T, Asahi T. Mechanically responsive crystals by light and heat. In: Koshima H, editor. Mechanically Responsive Materials for Soft Robotics. New York: Wiley; 2019. pp. 57-87. DOI: 10.1002/9783527 822201.ch3

[2] Inamura T, Hosoda H, Wakashima K, Miyazaki S. Anisotropy and temperature dependence of Young's modulus in textured TiNbAl biomedical shape memory alloy. Materials transactions. 2005;**46**:1597-1603. DOI: 10.2320/ matertrans.46.1597

[3] Namazu T, Hashizume A, Inoue S. Thermomechanical tensile characterization of NiTi shape memory alloy films for design of MEMS actuator. Sensors and Actuators A: Physical. 2007; **139**:178-186. DOI: 10.1016/j.sna.2006. 10.047

[4] Taniguchi T, Asahi T, Koshima H. Photomechanical azobenzene crystals. Crystals. 2019;**9**:437. DOI: 10.3390/ cryst9090437

[5] Wang C, Sun CC. The landscape of mechanical properties of molecular crystals. CrystEngComm. 2020;**22**: 1149-1153. DOI: 10.1039/C9CE01874C

[6] Desiraju GR. Supramolecular synthons in crystal engineering—A new organic synthesis. Angewandte Chemie International Edition. 1995;**34**:2311-2327. DOI: 10.1002/anie.199523111

[7] Kato M, Ito H, Hasegawa M, Ishii K. Soft crystals: Flexible response systems with high structural order. Chemistry–A European Journal. 2019;**25**:5105-5112. DOI: 10.1002/chem.201805641

[8] Kato M, Ishii K, editors. Soft crystals: Flexible response systems with high

structural order. Singapore: Springer Nature; 2023. p. 265. DOI: 10.1007/ 978-981-99-0260-6

[9] Naumov P, Chizhik S, Panda MK, Nath NK, Boldyreva E. Mechanically responsive molecular crystals. Chemical Reviews. 2015;**115**:12440-12490. DOI: 10.1021/acs.chemrev.5b00398

[10] Naumov P, Karothu DP, Ahmed E, Catalano L, Commins P, Mahmoud Halabi J, et al. The rise of the dynamic crystals. Journal of the American Chemical Society. 2020;**142**:13256-13272. DOI: 10.1021/jacs.0c05440

[11] Kitagawa D, Nishi H, Kobatake S. Photoinduced twisting of a photochromic diarylethene crystal. Angewandte Chemie. 2013;**125**:9490-9492. DOI: 10.1002/ ange.201304670

[12] Kitagawa D, Tsujioka H, Tong F, Dong X, Bardeen CJ, Kobatake S. Control of photomechanical crystal twisting by illumination direction. Journal of the American Chemical Society. 2018;**140**: 4208-4212. DOI: 10.1021/jacs.7b13605

[13] Al-Kaysi RO, Tong F, Al-Haidar M, Zhu L, Bardeen CJ. Highly branched photomechanical crystals. Chemical Communications. 2017;**53**:2622-2625. DOI: 10.1039/C6CC08999B

[14] Tong F, Xu W, Guo T, Lui BF, Hayward RC, Palffy-Muhoray P, et al. Photomechanical molecular crystals and nanowire assemblies based on the [2+ 2] photodimerization of a phenylbutadiene derivative. Journal of Materials Chemistry C. 2020;**8**:5036-5044. DOI: 10.1039/C9TC06946A

[15] Taniguchi T, Fujisawa J, Shiro M, Koshima H, Asahi T. Mechanical motion

of chiral azobenzene crystals with twisting upon photoirradiation. Chemistry–a. European Journal. 2016;**22**: 7950-7958. DOI: 10.1002/chem.2015 05149

[16] Hagiwara Y, Taniguchi T, Asahi T, Koshima H. Crystal actuator based on a thermal phase transition and photothermal effect. Journal of Materials Chemistry C. 2020;**8**:4876-4884. DOI: 10.1039/D0TC00007H

[17] Taniguchi T, Sugiyama H, Uekusa H, Shiro M, Asahi T, Koshima H. Walking and rolling of crystals induced thermally by phase transition. Nature Communications. 2018;**9**:538. DOI: 10.1038/s41467-017-02549-2

[18] Ishizaki K, Sugimoto R, Hagiwara Y, Koshima H, Taniguchi T, Asahi T. Actuation performance of a photo-bending crystal modeled by machine learning-based regression. CrystEngComm. 2021;**23**:5839-5847. DOI: 10.1039/D1CE00208B

[19] Taniguchi T, Kubota A, Moritoki T, Asahi T, Koshima H. Two-step photomechanical motion of a dibenzobarrelene crystal. RSC advances. 2018;**8**:34314-34320. DOI: 10.1039/ C8RA06639F

[20] Ahmed E, Karothu DP, Naumov P. Crystal adaptronics: Mechanically reconfigurable elastic and superelastic molecular crystals. Angewandte Chemie International Edition. 2018;**57**: 8837-8846. DOI: 10.1002/anie.2018 00137

[21] Worthy A, Grosjean A, Pfrunder MC, Xu Y, Yan C, Edwards G, et al. Atomic resolution of structural changes in elastic crystals of copper (II) acetylacetonate. Nature Chemistry. 2018;**10**:65-69. DOI: 10.1038/nchem. 2848

[22] Thompson AJ, Orué AIC, Nair AJ, Price JR, McMurtrie J, Clegg JK. Elastically flexible molecular crystals. Chemical Society Reviews. 2021;**50**: 11725-11740. DOI: 10.1039/D1CS 00469G

[23] Annadhasan M, Basak S, Chandrasekhar N, Chandrasekar R. Next-generation organic photonics: The emergence of flexible crystal optical waveguides. Advanced Optical Materials. 2020;**8**:2000959. DOI: 10.1002/adom.202000959

[24] Hayashi S, Koizumi T. Mechanically induced shaping of organic single crystals: Facile fabrication of fluorescent and elastic crystal fibers. Chemistry–A European Journal. 2018;**24**:8507-8512. DOI: 10.1002/chem.201801965

[25] Hayashi S, Koizumi T. Elastic organic crystals of a fluorescent π-conjugated molecule. Angewandte Chemie. 2016;**128**:2751-2754. DOI: 10.1002/ange.201509319

[26] Hayashi S, Ishiwari F, Fukushima T, Mikage S, Imamura Y, Tashiro M, et al. Anisotropic Poisson effect and deformation-induced fluorescence change of elastic 9, 10-Dibromoanthracene single crystals. Angewandte Chemie International Edition. 2020;**59**:16195-16201. DOI: 10.1002/anie.202006474

[27] Das S, Mondal A, Reddy CM. Harnessing molecular rotations in plastic crystals: A holistic view for crystal engineering of adaptive soft materials. Chemical Society Reviews. 2020;**49**: 8878-8896. DOI: 10.1039/D0CS00475H

[28] Das S, Saha S, Sahu M, Mondal A, Reddy CM. Temperature-reliant dynamic properties and Elasto-plastic to plastic crystal (rotator) phase transition in a metal oxyacid salt. Angewandte

Chemie International Edition. 2022;**61**: e202115359. DOI: 10.1002/ anie.202115359

[29] Mondal A, Bhattacharya B, Das S, Bhunia S, Chowdhury R, Dey S, et al. Metal-like ductility in organic plastic crystals: Role of molecular shape and dihydrogen bonding interactions in Aminoboranes. Angewandte Chemie International Edition. 2020;**59**: 10971-10980. DOI: 10.1002/ anie.202001060

[30] Saha S, Mishra MK, Reddy CM, Desiraju GR. From molecules to interactions to crystal engineering: Mechanical properties of organic solids. Accounts of Chemical Research. 2018;**51**: 2957-2967. DOI: 10.1021/acs. accounts.8b00425

[31] Takamizawa S, Miyamoto Y. Superelastic organic crystals. Angewandte Chemie. 2014;**126**: 7090-7093. DOI: 10.1002/ ange.201311014

[32] Pieczyska E, Gadaj S, Nowacki WK, Hoshio K, Makino Y, Tobushi H. Characteristics of energy storage and dissipation in TiNi shape memory alloy. Science and Technology of Advanced Materials. 2005;**6**:889-894. DOI: 10.1016/j.stam.2005.07.008

[33] Takamizawa S, Takasaki Y. Superelastic shape recovery of mechanically twinned 3, 5-Difluorobenzoic acid crystals. Angewandte Chemie. 2015;**127**: 4897-4899. DOI: 10.1002/ ange.201411447

[34] Takasaki Y, Sasaki T, Takamizawa S. Temperature-diversified anisotropic superelasticity and ferroelasticity in a 3-methyl-4-nitrobenzoic acid crystal. Crystal Growth & Design. 2020;**20**: 6211-6216. DOI: 10.1021/acs.cgd.0c00964

[35] Mutai T, Sasaki T, Sakamoto S, Yoshikawa I, Houjou H, Takamizawa S. A superelastochromic crystal. Nature Communications. 2020;**11**:1824. DOI: 10.1038/s41467-020-15663-5

[36] Sasaki T, Sakamoto S, Engel ER, Takamizawa S. An Organosuperelastic mechanism with bending molecular chain bundles. Crystal Growth & Design. 2021;**21**:2920-2924. DOI: 10.1021/acs.cgd.1c00089

[37] Sakamoto S, Sasaki T, Sato-Tomita A, Takamizawa S. Shape Rememorization of an Organosuperelastic crystal through Superelasticity–Ferroelasticity interconversion. Angewandte Chemie. 2019;**131**:13860-13864. DOI: 10.1002/ ange.201905769

[38] Sasaki T, Nishizawa K, Takamizawa S. Versatile Organosuperelastic deformability by multiple mechanical twinning. Crystal Growth & Design. 2021;**21**:2453-2458. DOI: 10.1021/acs.cgd.1c00054

[39] Sasaki T, Ranjan S, Takamizawa S. Perpendicularly oriented dual Organosuperelasticity correlated with molecular symmetry. Crystal Growth & Design. 2021;**21**:3902-3907. DOI: 10.1021/acs.cgd.1c00210

[40] Sasaki T, Ranjan S, Takamizawa S. A photoluminescent organosuperelastic crystal of 7-amino-4-methylcoumarin. CrystEngComm. 2021;**23**:5801-5804. DOI: 10.1039/D1CE00359C

[41] Sasaki T. Mechanical twinning in organic crystals. CrystEngComm. 2022; **24**:2527-2541. DOI: 10.1039/D2CE 00089J

[42] Mir SH, Takasaki Y, Takamizawa S. An organoferroelasticity driven by molecular conformational change.

Physical Chemistry Chemical Physics. 2018;**20**:4631-4635. DOI: 10.1039/C7CP07206F

[43] Seki T, Feng C, Kashiyama K, Sakamoto S, Takasaki Y, Sasaki T, et al. Photoluminescent ferroelastic molecular crystals. Angewandte Chemie International Edition. 2020;**59**: 8839-8843. DOI: 10.1002/anie.201914610

[44] Taniguchi T, Ishizaki K, Takagi D, Nishimura K, Shigemune H, Kuramochi M, et al. Superelasticity of a photo-actuating chiral salicylideneamine crystal. Communications Chemistry. 2022;**5**:4. DOI: 10.1038/s42004-021-00618-8

[45] Taniguchi T, Sato H, Hagiwara Y, Asahi T, Koshima H. Photo-triggered phase transition of a crystal. Communications Chemistry. 2019;**2**:19. DOI: 10.1038/s42004-019-0121-8

[46] Karothu DP, Weston J, Desta IT, Naumov P. Shape-memory and self-healing effects in mechanosalient molecular crystals. Journal of the American Chemical Society. 2016;**138**: 13298-13306. DOI: 10.1021/jacs.6b07406

[47] Ahmed E, Karothu DP, Warren M, Naumov P. Shape-memory effects in molecular crystals. Nature Communications. 2019;**10**:3723. DOI: 10.1038/s41467-019-11612-z

[48] Commins P, Hara H, Naumov P. Self-healing molecular crystals. Angewandte Chemie International Edition. 2016;**55**:13028-13032. DOI: 10.1002/anie.201606003

[49] Bhunia S, Chandel S, Karan SK, Dey S, Tiwari A, Das S, et al. Autonomous self-repair in piezoelectric molecular crystals. Science. 2021;**373**: 321-327. DOI: 10.1126/science.abg3886

[50] Park SK, Sun H, Chung H, Patel BB, Zhang F, Davies DW, et al. Super- and ferroelastic organic semiconductors for ultraflexible single-crystal electronics. Angewandte Chemie. 2020;**132**: 13104-13112. DOI: 10.1002/ange.202004083

[51] Sun H, Park SK, Diao Y, Kvam EP, Zhao K. Molecular mechanisms of superelasticity and ferroelasticity in organic semiconductor crystals. Chemistry of Materials. 2021;**33**: 1883-1892. DOI: 10.1021/acs.chemmater.1c00080

[52] Zhang N, Zaeem MA. Superelasticity and shape memory effect in zirconia nanoparticles. Extreme Mechanics Letters. 2021;**46**:101301. DOI: 10.1016/j.eml.2021.101301

[53] Ito S. Recent advances in mechanochromic luminescence of organic crystalline compounds. Chemistry Letters. 2021;**50**:649-660. DOI: 10.1246/cl.200874

[54] Takagi D, Ishizaki K, Asahi T, Taniguchi T. Molecular screening for solid–solid phase transition by machine learning. ChemRxiv. DOI: 10.26434/chemrxiv-2022-8z976-v2. [Preprint]

[55] Moriwaki H, Tian YS, Kawashita N, Takagi T. Mordred: a molecular descriptor calculator. Journal of Cheminformatics. 2018;**10**:1-14. DOI: 10.1186/s13321-018-0258-y

[56] Github. Available from: https://github.com/takuyhaa/Superelasticity-manifold/

[57] Butler KT, Davies DW, Cartwright H, Isayev O, Walsh A. Machine learning for molecular and materials science. Nature. 2018;**559**: 547-555. DOI: 10.1038/s41586-018-0337-2

[58] Himanen L, Geurts A, Foster AS, Rinke P. Data-driven materials science: Status, challenges, and perspectives. Advanced Science. 2019;**6**:1900808. DOI: 10.1002/advs.201900808

[59] Schmidt J, Marques MR, Botti S, Marques MA. Recent advances and applications of machine learning in solid-state materials science. npj Computational Materials. 2019;**5**:83. DOI: 10.1038/s41524-019-0221-0

[60] Musil F, De S, Yang J, Campbell JE, Day GM, Ceriotti M. Machine learning for the structure–energy–property landscapes of molecular crystals. Chemical Science. 2018;**9**:1289-1300. DOI: 10.1039/C7SC04665K

Chapter 5

A Review on Present Status of Friction Stir Welding of NiTinol, a Functionally Advanced, Versatile and Widely Used Shape Memory Alloy

Susmita Datta and Pankaj Biswas

Abstract

Ni-Ti alloys are extensively utilized in different fields of manufacturing because of their typical pseudoelastic effect and the shape memory properties. Welding of NiTinol is always essential to manufacture diverse geometrical structures with sufficient design flexibility following the application necessities. NiTinol is susceptible to compositional variations and microstructural changes because of the welding process. As a result, the mechanical and microstructural properties along with other functional properties get deteriorated with time. Welding of NiTinol without melting is extremely substantial because of the avoidance of the volatilization of the compositional constituents. Friction stir welding, a solid-state welding method, satisfies all the vital necessities of NiTinol alloy welding. This chapter will describe the friction stir welding of NiTinol in both similar and dissimilar material combinations in detail. The effect of different welding process variables on mechanical and metallurgical properties will be described along with the description of the smart functionality of the welded structures and the corrosion resistance performance.

Keywords: friction stir welding, NiTinol, metallurgy, mechanical properties, smart materials, shape memory alloys

1. Introduction

In today's modern engineering world, material science has progressed to develop unique and functional materials with advanced and strong structural performance and functionalities. The shape memory materials with smart functionalities fall beneath the wide variety of innovative engineering materials with superior and unique properties to identify and react to some certain stimulation by altering their chemical or physical properties. Shape memory alloys (SMA) can remember their original shape and can come back to it when thermal stimulation is applied to the deformed shape. SMA shows higher activation energy density and can recuperate to its original shape under high amounts of applied loads in comparison with other smart materials [1, 2]. Depending on the constituent components, SMAs can be characterized as NiTi-based (NiTi, NiTiZr, NiTiHf, NiTiCu), iron-based (FeNiCoTi, FeMnSi) and copper-based alloys (CuZnAl, CuAlBe, CuAlMn) [3–6].

SMAs have different crystal structures of different phases and the properties of different phases vary considerably. NiTinol has two phases. A steady phase at high temperature with a cubic structure is identified as austenite phase. The product phase at low temperature with monoclinic structure is known as martensite phase. The solid-state reversible phase transformation because of the shear lattice deformation in place of diffusion of atoms is the cause of the distinctive functional properties of SMA, such as pseudoelasticity (PE) and shape memory effect (SME).

The recovery of the shape by application of thermal stimulation is acknowledged as SME. NiTi is in the twinned martensite phase at room temperature (point 1). After solicitation of load, the detwinning of NiTinol happens. The maximum and minimum stress needed to distort the SMA is known as detwinning finish stress (σf) and detwinning start stress (σs), correspondingly [1]. The detwinning phenomena prompt a recoverable distortion where the allied stress will be smaller than the plastic yield stress value of low-temperature unstable martensite phase. After the elimination of the load, the elastic recovery occurs (point 4 to point 5) by holding the detwinned martensite phase. The shape retrieval happens because of the conversion of detwinned martensite to austenite by heating and it is known as reverse conversion. The start and end of this conversion sequence are represented as point 6 and point 7 equivalents to austenite start (As) and austenite finish temperature (Af), correspondingly. At austenite finish temperature (point 7), NiTi is existing in the austenite phase. This transformation sequence is known as forward conversion. The transformation starts and finish temperature of twinned martensite from austenite is characterized by Ms. and Mf. This whole process occurred in a cycle and is known as one-way shape memory effect (OWSME) as the shape recovery was obtained by heating and detwinning mechanism of the twinned martensite by application of outside mechanical load. OWSME is extensively applied for a widespread use. On contrary to this, the SMA may also be accomplished to show cyclic mechanical properties by the utilization of cyclic heat load devoid of any necessities of outside mechanical load. In plain words, simply, the cyclic phase conversion between austenite and martensite phase only by application of thermal load is recognized as two-way shape memory effect (TWSME). Because of the lower amount of strain recovery and the requirements of training, the TWSME is less employed for different practical applications [7].

The strain reclamation characteristics determined by stress-generated martensite conversion at temperatures over the austenite finish temperature (Af) but below Md is known as the pseudoelastic effect. The SMA will recuperate to the initial shape by removing the load at temperatures over Af [8, 9]. Over Af temperature, the pseudoelastic transformation can be initiated and can be continued to develop by application of outside load which helps in the generation of the detwinned martensite phase. Preferably, after the elimination of load, the route converses and high-temperature steady austenite phase generates at no load state. Conversely, the pseudoelastic transformation cycle of SMA helps in stabilizing the stress or stress-generated martensite and lowers the amount of strain recovery. The amount of irrecoverable strain escalates after each pseudoelastic cycle because of generated lattice defects and dislocations [10–12]. By solicitation of load, the austenite state goes through elastic loading (1–2). The beginning of the conversion from the austenite phase to detwinned martensite commences and the SMA faces a high value of inelastic strains [2, 3]. σ^{Ms} and σ^{Mf} are the values of required stress for starting and finishing the forward conversion correspondingly. After the accomplishment of forward conversion, the enhancement of stress level helps in elastic piling (3–4) which is basically the elastic transformation of the detwinned martensite. After removal of the force, martensite undergoes elastic unloading [4, 5]. After reaching the start stress (σ^{As}) of the austenite phase, the converse conversion from martensite phase to austenite starts and the recovery of the ends at austenite finish stress (σ^{Af}) (5–6). The route from point 6 to point 1 is described as the elastic repossession of the parent austenite phase. The whole loading-unloading curve of an SMA forms a hysteresis loop, and the area inside the loop depicts the amount of heat dissolute during a cycle.

2. About NiTinol alloy

NiTinol is the widely utilized and most preferred SMA among design engineers because of its synergistic amalgamation of superior functional and mechanical properties, pseudoelasticity and shape memory effect. This alloy has very good density and high deformation recovery properties [13]. The thermal responsive shape change has helped the utilization of NiTinol as actuators and sensors in different fields of robotic and industrial applications. Biocompatibility is the major requisite of any materials to be used in biomedical applications. Biocompatibility means that the material should not show any harmful effects, including any inflammatory, toxic or allergic reaction) during the working period inside the human body. Any biomaterial has to face a very critical environment inside the human body. NiTinol, a functionally advanced biomaterial, has very good corrosion protection characteristics and biocompatibility than any other alloy used in biomedical fields [14, 15]. A skinny and long-lasting protective layer of titanium oxide (TiO, Ti_2O_3 and TiO_2) on the surface of NiTinol helps to ensure the good corrosion resistance property of NiTinol. A good combination of superior characteristics, such as biocompatibility, shape memory effect, pseudoelasticity and kink resistance, helps in the use of NiTinol in the biomedical industry. A few of the extensively utilized biomedical NiTinol components are cardiovascular stents, orthopedic implants (spacers, connectors, staples and plates), orthodontic braces, minimally invasive surgery devices, forceps and aortic pumps.

The stress-strain relationship of NiTinol is greatly comparable with human bones and tissues [16]. As a result of this, the biomechanical characteristics of NiTinol match with the human body. A resemblance in loading and unloading between NiTinol and human bone was observed. The recoverability of strain of steel is below 0.5% but it shows a value of 8% for NiTinol. Typically, NiTinol having pseudoelasticity and the austenite finish temperature lower than the body temperature is favored for biomedical implants.

Even the shape recovery property of NiTinol was also used for surgical devices and implants. The biomedical devices could be activated in the body by utilizing body heat or any other exterior heat source to advance the bone joining and or nominal invasive surgery, correspondingly.

The requirement of close compositional constitute variance [9] of NiTinol demands a high degree of care during the manufacturing of the alloy. The steps involved in the production of NiTinol are very similar to the steps utilized for the production of traditional metallic materials. Those include melting and different hot and cold working methods. Along with these, an additional stage of shape memory treatment is carried out. Maintenance of appropriate and uniform constitutional composition and purity must be done during the melting phase to ensure the appropriate properties of the alloy. Carbon and oxygen contamination must be barred to control the purity. Alumina and magnesia crucibles are not generally favored because of oxide contamination issues. Generally, graphite crucibles are reused multiple times as the repetition in use helps the development of NiTinol coating on the inner surface of the crucible which basically decreases the carbon exposure in subsequent stages. The vacuum induction furnace (VIM) along with vacuum arc remelting (VAR) and electromagnetic excitation is used in sequential melting and remelting stages to confirm the uniformity in the molten material during the melting phase.

Because of the low formability and machinability of NiTinol, the processing is done in between the temperature range of 700 and 950°C. If cold working is performed for wire drawing and in any other applications, the intermediary annealing process is used to counteract the work hardening.

Shape memory treatment shows a vital role in the production of NiTinol. Basically, two aims are fulfilled by shape memory treatment. They are: (i) regulation of phase conversion temperature and (ii) memorizing the preferred shape. The shape memory treatment is basically performed in the temperature range of 300°C and 500°C. The processing and manufacturing steps and conditions considerably disturb the conversion temperatures and the mechanical characteristics. The alloy can show anyone of pseudoelasticity or shape memory effect at room temperature depending on the manufacturing route and constitutional composition. Maintaining the shape memory behavior is also an acute need in the processing of NiTinol. The occurrence of any unwanted and brittle intermetallic phases (Ti_2Ni, Ti_2Ni_3, and Ti_3Ni_4) or inclusion of nitrogen, hydrogen and oxygen elements should be eliminated. The occurrence of these elements changes the superior characteristics severely and may be the reason for the embrittlement of the NiTinol devices. In that scenario, the pseudoelastic property or the actuation mechanism by stress will not be shown.

The edge of the NiTi is nearly perpendicular to the Ti-rich side, but the Ni-rich side reduces with the reduction in temperature, and at about 500°C the solubility drips to insignificant. The diffusional conversion can happen and different phases, like Ti_2Ni_3, $TiNi_3$ and Ti_3Ni_4 can be generated depending on the amount of constituent components, the aging time and the temperature. The degree of diffusional conversion in comparison with the rise in aging time and temperature is as follows: $Ti_3Ni_4 \rightarrow Ti_2Ni_3 \rightarrow TiNi_3$. The NiTi phase exhibits the B2 structure at room temperature. The compositional constituent's amount must be in a specific range for this structure. B2 structure transforms to BCC at a temperature of 1090°C. This phase is retained during furnace cooling and quenching up to room temperature. The phase conversion temperature plays a vital part in the determination of the shape memory alloys' applicability. A negligible change (even 0.1 at %) in the constitutional composition has a significant effect on phase conversion (more than 10°C) [5, 6]. The phase conversion temperature may be custom-made depending on the solicitation by varying the Ni amount or by precipitate formation. The martensite start temperature rests almost constant till 49.7 at % of Ni. Above this percentage of Ni, the Ms. temperature exhibited a reducing trend. In this region, the austenite finish temperature (Af) is more or less 30 K above Ms. temperature.

The equiatomic conformation can show the highest value of austenite finish temperature (Af) of 120°C. The springs, made of NiTinol and utilized for hot water regulators, have Ni amount of 50.5 at %. With 51% Ni content, NiTinol can show pseudoelasticity having an austenite finish temperature of 40°C [1]. The elemental constituent range for the B2 state is very slender beneath 700°C. Depending on the heat treatments, the alloy having Ni more than 50% helps the development of Ni-rich intermetallic phases such as Ti_2Ni_3 and Ti_3Ni_4. The generation of precipitate alters the phase conversion temperature. To regulate the phase change temperature, different elements such as vanadium, copper and chromium can be supplemented.

3. Inevitability and obligations for NiTinol welding

Fabrication of NiTinol complex products are challenging and costly affair because of the poor machinability and low formability of the alloy. As a result of this, welding is a vital method for the fabrication of NiTinol devices with adequate freedom in design. The welding of NiTinol is a fascinating work because good welding needs sufficient mechanical properties and preservation of superior properties (pseudo-elasticity and shape memory effect) of the alloy. If the shape memory property is not retained in the welded structure, then the activation by application of either stress or temperature becomes difficult and prevents the desired application.

Microstructure and the constitutional composition were varied during welding. These variations may affect the phase conversion temperature of NiTinol and must have an unwanted effect on the thermomechanical stability of the fabricated structure [11, 12]. If there is a considerable mismatch in the phase conversion temperature of the joint and the parent material, then the control and actuation of the joined structure turn out to be very problematic. Still, a considerable variation in phase conversion temperature is fascinating and can impart a functionally graded characteristic in the fabricated structure. Welded NiTinol joints have a wide range of applications depending on the necessity of pseudoelasticity or shape memory effect. Dissimilar material combinations with steel have well-known applications in petrochemical, nuclear and aerospace applications [17]. NiTinol was fabricated to a titanium structure for a notched and adaptive nozzle used in Boeing B-777 for reducing noise levels during landing and launch [18]. While launching, the temperature at the outlet will be higher to activate the NiTinol actuator. The NiTinol activating device helps to reduce the engine noise by acting as a protrusion which basically tugs the whole structure downwards. In the cruising state at high elevation, the outlet temperature reduces because of variations in the environment and the engine condition. At this point, the base structure made of titanium will perform as the bias and helps to boost the assembly back to its original form [19, 20].

In multi-way activation of NiTinol actuators, two SMA sheets with dissimilar phase conversion temperatures need to be fabricated. Here, welding is the best method to fabricate this type of structure. The fabricated structure can be at three distinct locations depending on the temperature variation. At normal temperature i.e., at room temperature (RT), the plates should be in a straight-line situation. With the increase in temperature, the plate, with austenite temperature ranging from 40 to 50°C, activates. By additional increase in temperature, the second SMA sheet will also trigger. The whole activation of the plate generates the third position. This system is extensively used for morphing the rotor blades in aircraft. In this technology, each blade is accomplished at a particular position separately and can be fabricated to a principal hub. Depending on the fluid flow and the thermal criteria, the blades will twist consequently and the subsequent mechanical yield from the rotor can be changed.

In concrete structures reinforced with NiTinol for seismic isolation, the NiTinol bars having a diameter of 3 mm and length of 446 mm were stressed till 7% strain was reached. Then those stressed bars were twisted into circles and the boundaries were welded utilizing the TIG welding method. The same type of rings was coupled with some constant gap among them to generate the ribbed cylinder structure. After that concrete cylinders with a diameter of 15 cm and height of 30 cm were built around the NiTinol reinforcement. This type of concrete structure has improved strength properties and a higher value of failure strain than the normal concrete structures without NiTinol reinforcement [21].

Resistance spot welding was successfully used to join cylindrical NiTinol wire with rectangular one for orthodontic applications [22]. The kind of orthodontic wires, made of NiTinol and used for tooth adjustment, improve the design flexibility by fabricating a small piece of NiTinol wire with a leveling wire. The small pieces of wires would perform as hooks to twist the elastics, stops or omega loops to help the dentist to make the adjustment in teeth. Welding is basically an easy, faster and cost-effective method to fabricate patient-specific hooks, loops to wire and rings of different dimensions.

Because of the significance of welding as a method of fabrication, there are numerous published literature and articles on welding of NiTinol [23–27]. The possibility of solid-state welding of NiTinol was studied in a few literatures. Most of the work was concentrated on different fusion welding techniques of NiTinol, mostly laser welding of NiTinol. In view of the unexceptional benefits of the solid-state joining process for the fabrication of composition dependent alloys and dissimilar material combinations, this review work has been structured to talk about the friction stir joining of NiTinol alloy in detail.

4. Advantages of solid-state welding

Fusion joining of NiTinol was vastly reported by many researchers. Different fusion welding methods such as laser welding, tungsten inert gas welding, resistance welding and electron beam welding have been applied to join NiTinol [23, 24, 26]. The fusion welding of NiTinol undergoes different disadvantages in the generation of different intermetallic phases, favored vaporization of some elements. The solid-state fabrication of NiTinol was not studied comprehensively in comparison with different fusion welding methods.

Fusion welding of Nitinol has been comprehensively studied. Various synergistic arrangements of temperatures and pressures are used to join materials together. Here, no melting of materials takes place. The defects related to solidification, such as cracks and porosities, can be prevented significantly as melting of the material is not occurring. The atomic level bonding at the contact surface of the two materials is formed by heavy plastic deformation and high temperature during the solid-state welding method. Generally, filler materials are not used. In many cases, the interlayers of the lesser melting point material are utilized to help in the welding process. As the molten state of the material is prevented by solid-state welding, the generation of different brittle intermetallic compounds, which basically deteriorates the mechanical characteristics of the joint, can be avoided. Solid-state welding produces a joint with good dimensional control. The generated residual stress is less than any fusion welding method. The conservation of shape memory effect in the welded samples is far better for solid-state welding than soldering, brazing and different fusion welding methods. There are different categories of solid-state welding depending on the principle of plastic deformation and heat generation. They are impact-based processes, diffusion-based processes and friction-based processes. Here we will discuss about the friction stir welding of NiTinol.

Now a day, friction stir welding (FSW) is deliberated as a powerful method of joining materials in solid-state after its invention in the year 1991 by the Welding Institute (TWI) [28–30] for resolving the issues related to welding of aerospace grade aluminum alloys which were considered as non-weldable at that time because of the development of porosity in the weld-bead, reduced mechanical properties and differences in microstructure across different zones of the welding [31–39]. From that point of time, FSW was moving faster as a feasible welding method for a wide variety of alloys and metals to be used for different applications starting from space shuttles to microelectromechanical systems (MEMS) [40].

5. Friction stir welding

In FSW, a non-consumable and external rotational tool is utilized to do the welding in a solid-state. The tool mainly contains a pin (probe) and a shoulder. The diameter ratio between the pin (smaller) and the shoulder (larger) basically depends on the type and the thickness of the material to be welded [41] and in a few cases on the tool material as well [42]. During the FSW process, the tool revolves at a predetermined rate and also plunges into the workpiece material till the full depth of penetration is achieved by the application of pressure from the shoulder of the tool to the top surface of the workpiece. This will generate heat and cause the material softening around the tool by severe plastic deformation. At this point, the tool starts moving forward with a pre-set value (mm/min) along the line of the joint in solid-state. During processing, the plastically deformed softened material about the tool is relocated from the advancing side (where tool rotation and direction of travel are the same) to the retreating side (where tool rotation and direction of the travel are opposite to each other). The ratio between tool rotational speed and the welding speed has a considerable effect on the formation of joint area. After completion of the welding, the tool leaves the workpiece with a keyhole which is one of the distinguishing characteristics of FSW. In the initial phase, the revolving tool starts plunging on the meeting surfaces. In the second stage, the plunging gets completed before starting of the tool traverse along the joint interface. In the third stage, welding gets completed and extraction of the tool happens.

5.1 Friction stir welding of NiTinol

In the year 2017, the feasibility of FSW of NiTinol was demonstrated by Mani Prabu et al. [43]. Consequently, in-depth experimentation on the effect of tool rotating speed on different aspects of FSW-ed NiTinol was described by them [44]. The tool made of Densimet with a normal cylindrical probe was utilized for the experimentation [45]. The marks on the welding top surface and root face undoubtedly portrayed different phases of FSW like dwelling/ plunging phase, welding with traverse movement and retraction of the tool [46]. Severe plastic deformation and high temperature helped to refine grains through dynamic recrystallization phenomena. No detrimental intermetallic phases were observed in the joint nugget. The variation of phase conversion temperature after FSW was marginal in comparison with fusion welding methods. The welding done at 800 rpm had the least effect on phase transformation temperature in comparison with all other welds made at upper tool rotational speed. Moreover, a minor deviation in phase conversion temperature was observed across the different zones of the welding like stir zone, retreating side and advancing side [44]. This change in phase conversion temperature was caused by the variances in residual stress, grain size and dislocation density. The welding done at 1000 rpm revealed 17% higher yield stress in contrast with the base samples and the tensile strength value was about 66% of the base material. The tensile characteristics of the welding carried out at 1200 rpm were reduced because of the presence of the tool fragments inside the weld. The temperature dissemination during FSW of NiTinol was simulated using Comsol Multiphysics software on the finite element analysis platform. The boundary on the probe has lodged the extreme temperature on the advancing side. The forward velocity component and the tangential component of the same acts in the same way on the advancing side. As a result of this, the advancing side has displayed comparatively more temperature in comparison with the retreating side. The temperature was augmented linearly with the escalation in tool rotating speed because of the increase of heat produced by friction and excitation of the weld zone.

Deng et al. performed the welding of Nitinol with the help of the W–Re tool having a cylindrical probe. The overall cross-section illustrating diverse zones of the friction stir welded NiTinol was revealed by them. The stir zone consisted of refined and recrystallized grain. However, the HAZ and TMAZ consisted of refined and elongated grains because of the difference in plastic deformation and temperature across different areas. The overall tensile property of the joint after welding was reduced by the creation of different defects like kissing bond, lamellar structure, tunnel defects, tool inclusions ($W_{13}Re_7$), intermetallic compounds (Ti_2Ni) and tunnel defects. The post-weld pickling (PWP) treatments were done to eliminate the brittleness of the joint. The PWP carried out at 600 rpm has significantly improved the tensile property and the ultimate tensile strength value reached is 751 MPa, which is 79.1% of that of the parent material.

Abdollah Bahador et al. used a tungsten carbide tool with a plain cylindrical pin for welding of NiTinol [47]. The weld texture was examined with the help of EBSD analysis. It was found that refined and equiaxed grains form the entire welded region. The grain size was decreased from 49 μm of base material to 6.6 μm of the welded samples. Because of the inborn asymmetry of temperature and strain rate at diverse zones of FSW, the diverse areas of the joint had dissimilar and inhomogeneous microstructure initiated by diverse degrees of recovery and recrystallization through the weld. Likewise, the variation in tool rotating speeds helps in the creation of different textures. The welded specimen shows an improved yield strength value than the parent samples because of texture formation, spreading of tool wear elements in the joint and grain refinement. The welding done at 350 rpm showed the maximum yield stress and maximum fracture stress of value 765 MPa and 870 MPa consequently.

5.2 Corrosion resistance performance of friction stir welded NiTinol samples

The corrosion protection performance of friction stir welded samples was measured in 3.5% NaCl solution by potentiodynamic polarization (PDP) test, an electrochemical method of testing corrosion [45]. The overall breakdown potential was decreased to a lower value after welding. However, the current density increased after welding which considerably improved the deterioration of the corrosion resistance property. The amount of corrosion of the parent material and the welding at 1000 rpm were 4.97×10^4 mm/year and 1.62×10^3 mm/year consequently. It is well known that the parent material had a homogeneous grain structure but the FSW-ed samples had certain gradients in the grain distribution because of strain rate and temperature variation across diverse zones of the weld. Generally, the pin-affected area has the finest grains than other adjacent areas. Because of this, the generation of homogeneous and inactive oxide layer was prevented and caused in lowering the corrosion resistance performance of the welded samples. Even the variation in residual stress across different zones along with the inclusions of wear particles from the tool were the reasons behind the degradation of corrosion protection performance value. Parker et al. [48] also reported that the overall corrosion protection performance of the friction stir welded samples was depreciated in contrast with the parent material.

5.3 Functional properties of FSW-ed NiTinol samples

The damping characteristics of the FSW-ed samples were studied by using a dynamic mechanical analysis (DMA) study. Though the damping capability of the welded samples are bit lower than the parent samples, all welded samples exhibited good damping capabilities in the verified frequency value of (1, 10 and 20 Hz.) [49]. Even the cyclic load-deformation behavior for FSW-ed NiTinol was tested using a tensile cycle loading test. The maximum value of tensile stress was 320 MPa, 495 MPa and 590 MPa at strain levels of 2.5%, 3.5% and 4.5% [45].

Likewise, Abdollah Bahador et al. examined the cyclic loading characteristics under tensile load for FSW-ed NiTinol samples [47]. They reported deterioration of the cyclic behavior because of the occurrence of the twisted and increased number of dislocations, tool fragment particles in the weld material, high Schmid factor and high texturing. Even they reported the breakage of a few samples after a few number of tensile cycles.

The actuation characteristics of the FSW-ed NiTinol samples under diverse actuating methods such as electrical heating, laser heating and hot plate heating were studied by Mani Prabu et al. [44]. The recovery of the bending mechanism was utilized for the hot plate method. The specimen was set to a predefined U-shape putting it in an ice immersion and then placed it in a hot plate at a temperature of 65°C. After putting the sample on the hot plate, it completely recovered the generated stress and went back to the parent shape. The specimen has regained the parent shape by recovering the induced strain within 27 seconds.

The cyclic load-deformation behavior under electrical actuation was also studied by Mani Prabu et al. [45]. The welded sample was kept in a cantilever arrangement with some predefined load on the free side and then the specimen was stimulated for up to 300 cycles devoiding of any noteworthy deterioration of the displacement behavior. This electrical activation is economic and gives improved regulation of the actuation of SMA. Still, there is some possibility of destruction of the electrical contact during processing because of the halted actuation by contact mode.

Even the laser actuation method was also studied for achieving better displacements. The laser activation techniques offered noncontact heating mode and as a result, no connectors and wiring were required. Improved actuator movement was achieved by laser actuation rather than electrical actuation. The FSW-ed strips were actuated in a cantilever arrangement with the help of the laser heating method [49]. Laser actuation offers better actuation capability and a higher value of displacement than electrical heating because of the produced high temperature and impulsive nature of heating.

5.4 Dissimilar FSW of NiTinol

Deng et al. made the welding between NiTinol sheets and Ti6Al4V sheets with the help of the FSW technique [50]. They have used W-Re tool having a cylindrical probe with a tapered cross-section. In the initial stage, they have found the crack generation and tunnel defect formation in the nugget area because of the lower fluidity of the material and internal residual generation. To avoid this defect formation, they have preheated the back plate along with the base material up to 200°C. The preheating helped to form good joining by increasing the fluidity of the material together with the reduction in cooling rate and temperature gradient. The macrostructural cross-section of the welding at a rotating speed of 475 rpm and traverse speed of 23.5 mm/min, 30 mm/min and 60 mm/min consequently were analyzed. It was observed that onion rings were formed by material interweaving technique at 30 mm/min welding speed. The micrograph images showed the occurrence of lamellar construction in all the welded samples. The tunnel formation and kissing bond defect formation were reported at the interface of the weld at traverse speeds of 60 mm/min and 30 mm/min because of a reduction in metallic fluidity. While welding, the NiTinol moved towards the Ti6Al4V plate and formed the landmasses of Ti_2Ni in the matrix of Ti6Al4V. The joint got at a welding speed of 23.5 mm/min and showed an extreme value of tensile strength of 269 MPa.

Parker et al. made a joint between stainless steel and NiTinol with the help of W-Re tool having a tapered cylindrical probe [51]. The welding cross-sectional microstructure has shown a good mingling of the material in the stir zone. Along with this, various welding zones were noticeably observable. A broader TMAZ and HAZ were observed in the steel part. Contrarily, a narrower TMAZ with no HAZ was observed on NiTinol side. Because of grain size modification, the grains in stir zone get reduced and enhancement of microhardness was observed at the stir zone. The FSW-ed samples revealed the extreme tensile strength value of 705 MPa at normal temperature and 438 MPa at raised up temperature of 121°C. The welded samples showed a considerably low value of impedance in the electrochemical corrosion test during the corrosion behavior determination because of the non-uniformity of the oxide layer. The Tafel plot graphs for FSW-ed samples were studied. The corrosion current density and corrosion potential of the weld were moderately higher and lower in comparison with the base material consequently.

The composites use bulk NiTi ribbon for strengthening purposes as in the Al matrix was made by FSW techniques. The hybrid method as well as the conventional method supported by Joule heating was utilized in the fabrication of the composites. The high value of temperature because of the hybrid method of heating helped in enhanced material flow and caused high interfacial and tensile strength properties [52].

6. Summary of the study

In FSW, frictional force, generated by tool rotation is used for making the joint. Usually, this force produces heavy plastic deformation and heat at the interface of the welding. The dynamic loading is being facilitated in the FSW process and as a consequence of this, a high amount of strain was generated during welding. The properties and microstructure of the hot deformation process have similarities with the FSW process.

There are not many studies on the FSW of NiTinol. More researches are needed for a better understanding of the FSW of NiTinol. Works related to the effect of different tool profiles, tool material, welding speed, axial force, tool rotational speed and tool tilt angles should be tried to solve the issues related to the welding of NiTinol. This review work, shows that good quality joints of NiTinol can be made using FSW. The generated heat and cooling rate primarily control the microstructure in welding and generates grain size difference across different region of welding. Because of the grain size variation and inhomogeneity in microstructure across different regions in welding, the mechanical properties of welding structure get deteriorated. In FSW of NiTinol, a fine-grained microstructure across different areas of welding was obtained. As a result, the overall mechanical properties and the toughness of the joint were enhanced.

FSW is basically a pollution-less process as no toxic metal vapor and gas are generated. Porosity and hydrogen embrittlement were not formed during FSW. The heat-affected zone is very low and the overall mechanical and metallurgical properties are far better than any type of fusion welding process. The primary aim of this review work was to detect the advantages and disadvantages of FSW while joining the NiTinol. In comparison with any fusion welding method, FSW has many advantages like lesser heat input, improved mechanical and metallurgical properties, higher corrosion resistance property, lesser heat affected zone and no requirement of any special protective environment and filler material. FSW needed considerably less amount of power than any fusion welding method. The microhardness of FSW-ed NiTinol sample was very good. The microstructural, mechanical and metallurgical properties along with phase conversion temperature have a significant influence on different process parameters because of induced stresses and chemical composition variation during welding. The induced changes in FSW are minimal in comparison with any fusion welding method. As a result, FSW can be utilized for different applications where the phase conversion temperature of the welding should lie close to the parent sample. FSW helps to obtain enhanced weld strength because of substantial recrystallization phenomena achieved through severe plastic deformation.

This review on FSW of NiTinol shows that more work in similar and dissimilar material combinations is needed using traditional and hybrid FSW processes to acquire more knowledge in the field so that FSW could be a practical fabrication route for NiTinol components in aerospace, automotive, biomedical, hydrospace and civil structural application. Such research work would reduce the joining costs and help in the commercialization of the process for NiTinol fabrication.

The above work shows that most of the research work deals with the strength enhancement of the joint and metallurgical and mechanical aspects. In most of the research work the tool is made of expensive material. Such tools enhanced the overall cost of machining. It was observed that the generated microstructure in the biomaterial region was acceptable but the strength of the joint is not sufficient for industrial use for dissimilar NiTinol joints. At present different secondary heat sources were used for FSW of high-strength alloys to enhance the material flow around the tool and to get good quality joints at low heat input. So far, such studies with secondary heating have not reported for FSW of NiTinol. The research related to hybrid FSW of NiTinol with a secondary heat source needs more attention in the near future to commercialize the process for industrial adaptation of FSW of NiTinol at comparatively lower heat input.

7. Future scope of study

As FSW has exceptional proficiencies in joining NiTinol, a wide scope of future study is prominent. A few major research fields that should be explored in forthcoming years are listed here.

1. The FSW of NiTinol with Cu and Fe-based shape memory alloys should be studied as a dissimilar material combination with functional application because of the considerable difference in their phase conversion temperatures.

2. Simulation and modeling study of FSW is required for a better understanding of the effect of process parameters on weld qualities.

3. The consequence of FSW on fatigue property and the cyclic load-deformation behavior of NiTinol should be studied to know about the service life of the fabricated component.

4. Though a few corrosion studies of FSW-ed NiTinol in 3.5% NaCl solution was performed by previous researchers, the corrosion study in different physiological solution has not been explored yet.

5. The limited tool life because of the very high wear rate and the occurrence of tool rubbles in the weld zone is a serious problem that needs more research and attention to enhance tool life by reducing toll wear and preventing the tool debris particles to include in the welded region.

6. Hybrid FSW of the tool with some secondary heating source should be studied for industrial adaptation of FSW of NiTinol. These hybrid FSW processes reduce the requirement of overall heat input and enhance the material flow around the tool pin and helps in the formation of good-quality joint.

Acknowledgements

This work was supported by SERB, India. File Number: PDF/2021/001330.

Author details

Susmita Datta[1*] and Pankaj Biswas[2]

1 Indian Institute of Technology Guwahati, Guwahati, Assam, India

2 Department of Mechanical Engineering, IIT Guwahati, Assam, India

*Address all correspondence to: susmita.prod@gmail.com

IntechOpen

References

[1] Lagoudas DC. Shape Memory Alloys: Modelling and Engineering Applications. New York: Springer; 2008

[2] Sun L, Huang WM, Ding Z, Zhao Y, Wang CC, Purnawali H, et al. Stimulus-responsive shape memory materials: A review. Materials and Design. 2012;**33**:577-640. DOI: 10.1016/j.matdes.2011.04.065

[3] Oliveira JP, Shen J, Escobar JD, Salvador CAF, Schell N, Zhou N, et al. Laser welding of H-phase strengthened Ni-rich NiTi-20Zr high temperature shape memory alloy. Materials and Design. 2021;**202**:109533. DOI: 10.1016/j.matdes.2021.109533

[4] Oliveira JP, Schell N, Zhou N, Wood L, Benafan O. Laser welding of precipitation strengthened Ni-rich NiTiHf high temperature shape memory alloys: Microstructure and mechanical properties. Materials and Design. 2019;**162**:229-234. DOI: 10.1016/j.matdes.2018.11.053

[5] Oliveira JP, Panton B, Zeng Z, Omori T, Zhou Y, Miranda RM, et al. Laser welded superelastic Cu-Al-Mn shape memory alloy wires. Materials and Design. 2016;**90**:122-128. DOI: 10.1016/j.matdes.2015.10.125

[6] Oliveira JP, Zeng Z, Berveiller S, Bouscaud D, Braz Fernandes FM, Miranda RM, et al. Laser welding of Cu-Al-Be shape memory alloys: Microstructure and mechanical properties. Materials and Design. 2018;**148**:145-152. DOI: 10.1016/j.matdes.2018.03.066

[7] Mohd Jani J, Leary M, Subic A, Gibson MA. A review of shape memory alloy research, applications and

opportunities. Materials and Design. 2014;**56**:1078-1113

[8] Concilio A, Antonucci V, Auricchio F, Lecce L, Sacco E. Shape Memory Alloy Engineering: For Aerospace, Structural, and Biomedical Applications. First ed. Oxford, UK: Butter-Heinemann; 2015. DOI: 10.1016/B978-0-12-819264-1.01001-3

[9] Yamauchi K, Ohkata I, Tsuchiya K, Miyazaki S. Shape Memory and Superelastic Alloys Technologies and Applications. First ed. Cambridge, UK: Woodhead Publishing Limited; 2011. DOI: 10.1201/b16545

[10] Yoneyama T, Miyazaki S. Shape Memory Alloys for Biomedical Applications. Cambridge, UK: CRC Press; 2009. Available from: http://link.springer.com/10.1007/978-0-387-47685-8%0A; http://www.springer.com/series/8886%0A; http://link.springer.com/10.1007/978-3-319-031880%0A; DOI: 10.1016/j.jallcom.2014.12.009%0A; 10.1016/S1369-7021(07)70047-0%0A

[11] Oliveira JP, Miranda RM, Schell N, Fernandes FMB. High strain and long duration cycling behavior of laser welded NiTi sheets. International Journal of Fatigue. 2016;**83**:195-200. DOI: 10.1016/j.ijfatigue.2015.10.013

[12] Oliveira JP, Fernandes FMB, Schell N, Miranda RM. Martensite stabilization during superelastic cycling of laser welded NiTi plates. Materials Letters. 2016;**171**:273-276. DOI: 10.1016/j.matlet.2016.02.107

[13] Elahinia MH. Shape Memory Alloy Actuators: Design, Fabrication and Experimental Evaluation. West Sussex, UK: John Wiley & Sons, Ltd; 2016

[14] Machado LG, Machado LG, Savi M. Medical applications of shape memory alloys. Brazilian Journal of Medical and Biological Research. 2003;**36**:683-691

[15] Elahinia MH, Hashemi M, Tabesh M. Manufacturing and processing of NiTi implants: A Rev. Progress in Materials Science. 2012;**57**:911-946. DOI: 10.1016/j.pmatsci.2011.11.001

[16] Morgan NB. Medical shape memory alloy applications — the market and its products. Materials Science and Engineering A. 2004;**378**:16-23. DOI: 10.1016/j.msea.2003.10.326

[17] Tripathy I, Rout SP, Mallik M. Effect of temperature and pressure on diffusivity of nitinol pellet bonded with steel plate. Materials Today Proceedings. 2020;**33**:5213-5217. DOI: 10.1016/j.matpr.2020.02.892

[18] Chau ETF. A Comparative Study of Joining Methods for a SMART Aerospace Application. Cranfield University; 2007

[19] Friend PC, Allen PD, Webster J, Clark D, Goffin PK. Comparative Study of Joining Methods for a SMART Aerospace Application. Cranfield University; 2007

[20] Liu B, Wang Q, Hu S, Zhang W, Du C. On thermomechanical behaviors of the functional graded shape memory alloy composite for jet engine chevron. Journal of Intelligent Material Systems and Structures. 2018;**29**:2986-3005. DOI: 10.1177/1045389×18781257

[21] Choi E, Park S, Cho B, Hui D. Lateral reinforcement of welded SMA rings for reinforced concrete columns. Journal of Alloy Compound. 2013;**577**:S756-S759. DOI: 10.1016/j.jallcom.2012.02.135

[22] Mesquita TR, Martins LP, Martins RP. Welding strength of NiTi wires. Dental Press Journal of Orthotics. 2018;**23**:58-62. DOI: 10.1590/2177-6709.22.3.058-062.oar

[23] Oliveira JP, Miranda RM, Fernandes FMB. Welding and joining of NiTi shape memory alloys: A review. Progress in Materials Science. 2017;**88**:412-466. DOI: 10.1016/j.pmatsci.2017.04.008

[24] Kannan TDB, Ramesh T, Sathiya P. A review of similar and dissimilar micro-joining of nitinol. Journal of Minerals and Materials Society. 2016;**68**:1227-1245. DOI: 10.1007/s11837-016-1836-y

[25] Akselsen OM. Joining of shape memory alloy. In: Cismasiu C, editor. Shape Memory Alloys. Rijeka, Croatia: IntechOpen; 2004. pp. 183-211

[26] Mehrpouya M, Gisario A, Elahinia M. Laser welding of NiTi shape memory alloy: A review. Journal of Manufacturing Processes. 2018;**31**:162-186

[27] Mehta K, Gupta K. Fabrication and Processing of Shape Memory Alloys. Springer; 2019

[28] Thomas WM, Nicholas ED, Needham JC, Murch MG, Templesmith P, Dawes CJ. Friction Stir Welding. G.B. Patent Application No. 9125978. 1991

[29] Dawes C, Thomas W. Friction Stir Joining of Aluminium Alloys. Cambridge, UK: TWI Bulletin 6; TWI; 1995

[30] Thomas WM, Nicholas ED, Needham JC, Murch MG, Templesmith P, Dawes CJ. Friction Welding. U.S. Patent No. 5,460,317. 1995

[31] Mishra RS, Ma ZY. Friction stir welding and processing. Materials Science & Engineering R: Reports. 2005;**50**:1-78

[32] Çam G, Javaheri V, Heidarzadeh A. Advances in FSW and FSSW of dissimilar Al-alloy plates. Journal of Adhesion Science and Technology. 2022;**37**:162-194

[33] Kashaev N, Ventzke V, Çam G. Prospects of laser beam welding and friction stir welding processes for aluminum airframe structural applications. Journal of Manufacturing Processes. 2018;**36**:571-600

[34] Heidarzadeh A, Mironov S, Kaibyshev R, Çam G, Simar A, Gerlich A, et al. Friction stir welding/processing of metals and alloys: A comprehensive review on microstructural evolution. Progress in Materials Science. 2021;**117**:100752

[35] Çam G, ˙Ipeko˘glu G. Recent developments in joining of aluminum alloys. International Journal of Advanced Manufacturing Technology. 2016;**91**:1851-1866

[36] Çam G, Mistikoglu S. Recent developments in friction stir welding of al-Alloys. Journal of Materials Engineering and Performance. 2014;**23**:1936-1953

[37] Ipeko G, Çam G. Formation of weld defects in cold metal transfer arc welded 7075-T6 plates and its effect on joint performance. IOP Conference Series Materials Science and Engineering. 2019;**629**:012007

[38] Çam G. Prospects of producing aluminum parts by wire arc additive manufacturing (WAAM). Materials Today Proceedings. 2022;**62**:77-85

[39] Çam G, Koçak M. Microstructural and mechanical characterization of electron beam welded Al-alloy 7020. Journal of Materials Science. 2007;**42**:7154-7161

[40] Ahmed MMZ. The Development of Thick Section Welds and Ultra-Fine Grain Aluminium Using Friction Stir Welding and Processing. Sheffield, UK: The University of Sheffield; 2009

[41] Ahmed MMZZ, Wynne BP, Rainforth WM, Addison A, Martin JP, Threadgill PL. Effect of tool geometry and heat input on the hardness, grain structure, and crystallographic texture of thick-section friction stir-welded aluminium. Metallurgical and Materials Transactions A. 2018;**50**:271-284

[42] Ahmed MMZ, Wynne BP, Martin JP. Effect of friction stir welding speed on mechanical properties and microstructure of nickel based super alloy Inconel 718. Science and Technology of Welding and Joining. 2013;**18**:680-687

[43] Prabu SSM, Madhu HC, Perugu CS, Akash K, Kumar PA, Satish VK, et al. Microstructure, mechanical properties and shape memory behaviour of friction stir welded nitinol. Materials Science and Engineering A. 2017;**693**:233-236

[44] Prabu SSM, Madhu HC, Perugu CS, Akash K, Mithun R, Kumar PA, et al. Shape memory effect, temperature distribution and mechanical properties of friction stir welded nitinol. Journal of Alloy Compound. 2019;**776**:334-345. DOI: 10.1016/j.jallcom.2018.10.200

[45] Mani Prabu SS, Perugu CS, Madhu HC, Jangde A, Khan S, Jayachandran S, et al. Exploring the functional and corrosion behavior of friction stir welded NiTi shape memory alloy. Journal of Manufacturing Processes. 2019;**47**:119-128. DOI: 10.1016/j.jmapro.2019.09.017

[46] Deng H, Chen Y, Li S, Chen C, Zhang T, Xu M, et al. Microstructure, mechanical properties and

transformation behavior of friction stir welded Ni50.7Ti49.3 alloy. Materials Design. 2020;**189**:1-10. DOI: 10.1016/j.matdes.2020.108491

[47] Bahador A, Umeda J, Tsutsumi S, Hamzah E, Yusof F, Fujii H, et al. Asymmetric local strain, microstructure and superelasticity of friction stir welded Nitinol alloy. Materials Science and Engineering A. 2019;**767**:138344. DOI: 10.1016/j. msea.2019.138344

[48] West P, Shunmugasamy VC, Usman CA, Karaman I, Mansoor B. Part I.: Friction stir welding of equiatomic nickel titanium shape memory alloy – microstructure, mechanical and corrosion behavior. Journal of Advanced Joining Process. 2021;**4**:100071. DOI: 10.1016/j.jajp.2021.100071

[49] Prabu SSM, Palani IA. Investigations on the actuation behaviour of friction stir–welded nickel titanium shape memory alloy using continuous fibre laser. Journal of Microencapsulation. 2021;**5**(2):137-143. DOI: 10.1177/25165984211015409

[50] Deng H, Chen Y, Jia Y, Pang Y, Zhang T, Wang S, et al. Microstructure and mechanical properties of dissimilar NiTi/Ti6Al4V joints via back-heating assisted friction stir welding. Journal of Manufacturing Processes. 2021;**64**:379-391. DOI: 10.1016/j.jmapro.2021.01.024

[51] West P, Shunmugasamy VC, Usman CA, Karaman I, Mansoor B. Part II.: Dissimilar friction stir welding of nickel titanium shape memory alloy to stainless steel – microstructure, mechanical and corrosion behavior. Journal of Advanced Joining Process. 2021;**4**:100072. DOI: 10.1016/j.jajp.2021.100072

[52] Oliveira JP, Duarte JF, In'acio P, Schell N, Miranda RM, Santos TG. Production of Al/NiTi composites by friction stir welding assisted by electrical current. Materials and Design. 2017;**113**:311-318. DOI: 10.1016/j.matdes.2016.10.038